CULTURE, PLACE, AND NATURE
Studies in Anthropology and Environment
K. Sivaramakrishnan, Series Editor

Centered in anthropology, the Culture, Place, and Nature series encompasses new interdisciplinary social science research on environmental issues, focusing on the intersection of culture, ecology, and politics in global, national, and local contexts. Contributors to the series view environmental knowledge and issues from the multiple and often conflicting perspectives of various cultural systems.

Gardens of Gold

PLACE-MAKING IN PAPUA NEW GUINEA

Jamon Alex Halvaksz

UNIVERSITY OF WASHINGTON PRESS

Seattle

Gardens of Gold was supported by a grant from the Samuel and Althea Stroum Endowed Book Fund.

Composed in Warnock Pro, typeface designed by Robert Slimbach
Cover design by Katrina Noble
Cover photograph: James planting a yam. Photograph by the author.
All interior illustrations are by the author.

24 23 22 21 20 5 4 3 2 1

Printed and bound in the United States of America

UNIVERSITY OF WASHINGTON PRESS
uwapress.uw.edu

LIBRARY OF CONGRESS CATALOGING-IN-PUBLICATION DATA ON FILE

ISBN 978-0-295-74760-6 (hardback)
ISBN 978-0-295-74759-0 (paperback)
ISBN 978-0-295-74761-3 (ebook)

The paper used in this publication is acid free and meets the minimum requirements of American National Standard for Information Sciences—Permanence of Paper for Printed Library Materials, ANSI Z39.48–1984.∞

CONTENTS

FOREWORD

In this book we find, through their engagement with both gold mining and nature conservation, Biangai villagers articulating senses of place and expressing their attachments to territories replete with meaning and multigenerational investments. Yet this is not only landscape as myth or history, repository of tradition, or fount of ancestral knowledge. It is also the land from which new hopes and aspirations spring. For, as Jamon Halvaksz notes, land has duration and landscapes have appearances. Biangai emerge from this time-space topology, in his felicitous term, as "placepersons." In its dynamic approach to the making of places, interweaving remembrance with hope or aspiration, this work is alert to the longer history of colonialism and missionaries. And in it the Biangai are very aware of the more recent history of extractive industries, biological research, conservation industries, and ecotourism; gold and its mining do not, in their experience, convey unalloyed pleasure or unremitting pain.

The study is set in a part of the world that has a rich tradition of scholarship on these themes. Papua New Guinea has an abundant literature on place-making, the relations between places and human settlements, and the diversity of natural resources that lead to projects for extraction or conservation. As many of those prior studies have persuasively argued, Papua New Guineans have both resisted the forces of international capitalism with mixed results (Kirsch 2014) and found ways to join the commodity chains that inevitably emerged from their bountiful lands (West 2012). Halvaksz, based in Biangai villages along the Upper Bulolo River of Morobe Province, provides an account of extractive industries and commodity production that shows how places and personhood for the Biangai are profoundly shaped by diverse forms of production. Such production—be it agriculture, social relations forged through kinship and mobile emplacement in a larger Oceanic world, gold mining, coffee cultivation, or nature conservation—have had a dynamic and historical relation to Biangai ideas of place, heritage, and being human.

With these claims in hand, through a sustained and subtle ethnography that ranges across working the land in different ways, changing ideas about kinship, the outcomes of corporate mining projects, and large nature conservation endeavors, Halvaksz introduces and illuminates his analysis of placepersons. This concept is useful, as he shows, to describe the mutual constitution of places and people and thus offer a new and very promising direction for the more-than-human anthropology inspired by recent scholarship on indigenous worldviews and the natural environment in the Americas, Oceania, and Asia. Further, in this work, Halvaksz is keen to provide a historically situated account of the Biangai as authors of development narratives about the economic possibilities of what is found in the soil or in the depths below it. Farm produce, gold, and biodiversity are to be understood as the fruit of generations of place-making and the ways in which placepeople invoke and elicit virtue and reward from the land through their actions.

If Oceanic cultures constructed space-place relations through mobility across a broader region than the one in which they mostly resided, Halvaksz is keen to show that the Biangai also illuminated other ways of making places in an idiom familiar to modern ideas of development by describing their rootedness in territory and claims over generations to resources embedded in their territories. In this way, this book joins discussions on Oceania and indigenous studies even as it makes signal contributions to environmental anthropology and development studies.

The deep reading and engagement of indigenous scholarship is very beneficial to this study, for through it Halvaksz shows the connections Biangai make with and sustain in places in the Upper Bulolo Valley, even as they participate in crafting their commodity futures and industrial development and conservation in their ancestral lands. His study advances a systematic association between ideas of rootedness and dialogic practices of communication. Lands and soils give shape to conditions and opportunities; laws and investments, through technologies like neoliberal modes of development, also introduce classification and fixity. Yet regional traditions of mobility and exchange provide a welcoming climate for blending ideas from near and far into new kinds of production and emplacement. Thus, Biangai people embody dynamic rituals of becoming native in the midst of continuing foreign incursions. Halvaksz has written a wonderful book that points to alternate approaches we might take for blending ecological and social transformation into textured analysis.

K. SIVARAMAKRISHNAN
Yale University

PREFACE

This book is about the social production and reproduction of an indigenous landscape, and the way in which our relationships with places can provide hope for a more just and equitable world. In contrast to many accounts of mining areas (e.g., Kirsch 2006; F. Li 2015), and other resource frontiers (Cepek 2018; Grandia 2012; Kosek 2006; Powell 2018; Sawyer 2004), the communities at the heart of this ethnography desire resource extraction on lands that they still control. In their effort to seek out futures in the detritus of industrial gold mining, Biangai villages along the Upper Bulolo River of Morobe Province in Papua New Guinea maintain a tenuous hold on social relations through place. Belonging to the places of their ancestors is more than just a local practice. In Oceanic theories of place, places are folded into more-than-human socialities where the distinctions between place and person become blurred. This volume's consideration of Biangai world-making from the position of *placepersons* provides both an ethnographic description of Biangai practices and an intervention in development policies that continue to value alienable rights for capitalist investment. In agreement with indigenous scholarship on place-thought (Watts 2013), Oceanic theories of *tā-vā*/time-space relations (e.g., Ka'ili 2017; Māhina 2010), and the decolonization of methods and theory (e.g., L. Smith 1999; TallBear 2014), its focus on placepersons reflects the mutual creation of the physical landscape and human sociality rooted in Biangai worldviews. Amid complex entanglements of places and development, attachments through place can be a source of resiliency in what might otherwise be understood as a frontier of neoliberal capital.

While gold mining might challenge the resolve of place-based relationships in other locations, for Biangai gold fulfills the promise of mythohistorical ancestors to provide for their descendants. Whereas I personally view gold mining as inherently detrimental for subsistence, community stability, and long-term environmental sustainability, my Biangai friends and colleagues view it as a transformative substance. Companies *should* want to work their lands because of Biangai connections to places, and gold *should* emerge from the ground *because* Biangai continue to tend to their relationships through the land.

Mining is also an easy metaphor for describing postcolonial place relations along the Bulolo River. Perhaps it is because mining directly evokes the depth of the landscape, folding sediments of earth and history into the contemporary experiences of places and persons. Since the early 1920s, the Upper Bulolo Valley has been a mined place. A relatively long history of mineral exploitation continues to inform how Biangai make their worlds. Every living Biangai man and woman has grown up in a world where mining for gold not only is possible but conditions the wider social life of the community. As a result, when Biangai think of economic opportunity and development, they think of gold. Today, they willingly seek out mineral exploration on their own lands. Logging, coffee, small businesses, and even conservation are all possibilities, but a seam of gold remains ideal. While seemingly contrary to Western understandings of sustainability, indigenous rights, and healthy environments, mining has become a Biangai metaphor for development and modernity.

This is also an ethnography of agriculture and conservation, two practices that are rearranged in light of the Bulolo Valley's golden history. Coffee, which circulates beyond the boundaries of Papua New Guinea, is increasingly associated with gold. Agricultural officers and local coffee buyers refer to coffee cherries as red gold, emphasizing that unlike yellow gold, red gold returns every year. Conservation is likewise read through extractive metaphors, as Biangai exchange specimens, photos, and knowledge of the forest for wealth and relationships with an international cadre of researchers. While the practices of gardening, conservation, and mining bring labor into direct relationships with place, gold's higher value and desirability beyond the valley give it weight in contemporary discourses and Biangai political and ecological relationships. Through these world-making practices, indigenous landowners try to make development possible. Their insights into place offer political interventions on which others might model more sustainable relationships with the worlds around us.

ACKNOWLEDGMENTS

This book brings together several different research trips, shifts in my own residence, and affiliations with different universities. Weaving together place and text, I am grateful to the people who shared those moments with me. Perhaps most prominent are my friends and host families in Papua New Guinea. In Elauru, Joe and Nawasio Songoa treated me as one of their own, bringing me into a family made richer by my brothers and sisters: Paipe, Chris, Kabi, John, Aris, James, Jacob, and Marawa, as well as Moses, Yateng, and Kuiya. GK, Kawena, Nalu, and Kipas were the true friends I really needed, and Yansom, James, Yawa James, Kikingai, Sam, Ben, Nare, Kaia, Kule, Yambu, Awa, Ana, Robert, Wanoa, Pai, Taisen, Simani, Dennis, and Jerry all provided guidance at different points. I couldn't have done this without the knowledge shared by Ilau, Yalimu, Yansom, Givisa, Yatu, and Kausa. In Winima, Philip and Sabu Kagowe were always welcoming, and Kuskom, Sowi, Kausa, Yamu, and Lambio always treated me as a brother. Within the community, Koizi, Peter, Kaya K., Korua, Marowa, Yapu, Yandi, Karibi, Kency, Kasi, Mangisa, Pari, Kiwi, Douglas, Harold, and Daniel were always helpful. Ngawae Mitio, from the Biangai village of Werewere and faculty at Unitech in Lae, has always supported my work, providing valuable feedback and insight. I owe a special thanks to him as well as to Kawen, Rex, and Kiwi for their insight into Winima's relationships with mining.

At the University of Minnesota, Kathleen Barlow and David Lipset provided guidance, and David has continued to be a sounding board, reading drafts of the book manuscript and offering advice and support, as well as a drink or two. I am also grateful to Steven Gudeman and Bruce Braun for their help as I thought about the issues that eventually migrated to this text. The late Gene Ogan also offered guidance at key points along the way. While I'm sure to forget some names, I also have to thank David Weinlick, Sam Bullington, Kathleen Saunders, Solveig Brown, and Theresa Early, as well as

John Soderberg, Emily Weglian, Lynn Newton, Taku Sazuki, Amanda Swarr, Jennifer Stampe, Amy Porter, Eric Bangs, Silas Mallery, and Elizabeth Hochberg.

At the University of Canterbury (Christchurch), David Gegeo, Karen Nero, Moana Matthes, Keith Camacho, Malakai Koloamatangi, Silipa Silipa, Katerina Teaiwa, Konai Thaman, Eric Waddell, Steven Winduo, and Heather Young-Leslie all shaped my understanding of wider Oceanic scholarship in important ways. My fellowship there funded a trip to Papua New Guinea in 2005, as well as provided a creative space to write and workshop parts of chapters 4 and 7.

The *Contemporary Pacific* and the *Journal of the Royal Anthropological Society* published earlier versions of chapters 2 and 7, respectively. Versions of different chapters were presented at the annual conferences of the American Anthropological Association and the Association for Social Anthropology in Oceania. Both have been fruitful venues for thinking through the breadth of anthropological theory and practice. Invited presentations at the University of Kentucky and the Ecology and Culture University Seminar at Columbia University provided equally rich feedback.

I'd like to thank the University of Minnesota, the National Science Foundation (BCS-0003927), the Wenner Gren Foundation for Anthropological Research (2000–2002: GR6724; 2014–16: GR50638), the University of Canterbury at Christchurch, and the University of Texas at San Antonio for funding phases of my research.

Over the last twelve years, I've called San Antonio home. My colleagues here have been incredibly supportive, and I truly appreciate both their continued collegiality and their friendship, which is rare in academia. Michael Cepek, Jill Fleuriet, and Patrick Gallagher have been great sounding boards for ideas. Thad Bartlett, Jason Yeager, Kat Brown, Laura Levi, Robert Hard, Sonia Alconini, Carolyn Ehardt, and Luca Pozzi have always offered support for my work. I've also been lucky to have some absolutely brilliant students with whom I've shared texts, discussed the latest ethnography, or discussed writing, including Alex Antram, Jason Roberts, Marissa Shaver, Becky DelliCarpini, Jennifer Torpie-Sweterlitsch, Christina Fraiser, Richard Stout, Jess Reid, and Michelle Vryn.

The University of Washington Press has provided a great space to see this through. Thanks to the editorial staff and especially Lorri Hagman and Dr. Kalyanakrishnan Sivaramakrishnan, as well as two anonymous reviewers, who provided guidance in finalizing the text and gave it a place in their wonderful series Culture, Place and Nature.

My colleagues, as generous with their time as with their friendship, have kept me sane, especially Joshua Bell, Jerry Jacka, Paige West, and JC Salyer. I have to acknowledge as well the role of Stub and Herbs, bouncy chairs, Dunn Brothers, ASAO, Rosellas, Press, and noisy coffee shops everywhere.

I wouldn't have learned anything from my research if I didn't thank my parents and sisters, and acknowledge the unique role they play in how I see the world. In different ways, they showed me how connections, kinships, and places across Kentucky are made real.

My daughter, Margaret, has proven to be a great muse, who may not have always understood why I had to be away for research but was always supportive. Of course, all of these places were shared with my friend and partner for more than two decades, Rachel. I owe them both such a debt of gratitude for putting up with my travels and the odd schedule of an academic. They have both always been there for me, and I dedicate this work to them.

GARDENS OF GOLD

Introduction

Above the Alluvial Flats

KAWNA squatted in the shallow creek bed, raising himself high enough off of the ground to avoid dampening his shorts. He pulled the sleeves of his gray, hooded fleece past his elbows, and carefully dipped his gold pan into the rushing water to wash the soil and rock. It was early. The water was cold against the skin, awaking the senses to the day. Children were still bathing nearby for school, and families were making their way toward the day's work in the coffee gardens that line the edges of the streams and rivers along the valley floor. But they all gathered around Kawna as he skillfully rotated and examined the contents of his pan. Everyone waited in anticipation. The calm waters of the Ngayaka, which feeds into the Bulolo River, were blackened by the fine particles of soil. Set free by our movements, they flowed gently downstream toward the coast. We all stared hopefully into Kawna's dish. To find gold at that moment would be transformative but not unexpected. Biangai anticipate that fortune will eventually befall them.

On the way back from his sweet potato gardens the night before, Kawna had noticed that part of the steep rocky slope near the creek had slipped down, revealing what looked like a layer of sediment speckled with quartz gleaming in the evening light. "A sign of gold," he said to me that night as we made plans for the next day. In the morning, he loaded his gold pan into his gear as we set out together to cut posts of *kolonangai* (a tropical hardwood) for his new house. But first he wanted to try his hand at panning beneath the landslide. Geologists exploring Morobe Province for a multitude of mining companies have told Biangai that gold does exist deep beneath their lands, and Biangai have found gold in the river and creek beds around their villages. The location of Kawna's efforts was also auspicious in Biangai mythopoetic histories, for it was from the top of the high ridge rising above us that a great mythical warrior once tumbled. His fall has long haunted the site, informing how Biangai interpret their experiences of the place. As in other locations, such a spiritual presence lends weight to any prognosis of discovery of gold (Halvaksz and Young-Leslie 2008; see also Biersack 1999) and demonstrates the kinds of connections that Biangai make with a very real, living landscape. The dish, the value of gold, and the knowledge that brought us to the site were but mediators, which Kawna hoped would connect him to good fortune. It is such relations that inform this ethnography.

When it comes to development, Biangai speakers living in the Bulolo Valley (Morobe Province) have experience with cash cropping, logging, urban wage labor, small-scale and industrial mining, as well as conservation efforts driven by ecotourism and research. With colonial gold mining starting in the early 1920s, and missionaries and government officers following close behind, Biangai are not the stereotypical isolated population that characterizes popular imaginaries about Papua New Guinea (West 2016). While each of these experiences continues to inform and mark the landscape, this ethnography attends to the very specific regimes of conservation and industrial gold mining in two neighboring communities. The village of Winima maintains interest in an international industrial gold mine developed on land of which they claim part ownership. As a result, they receive various forms of compensation and employment opportunities. Elauru, where Kawna hails from, has a historical and problematic relationship with conservation and scientific research in the valley, and like all Biangai communities, they aspire to find gold on their land.

Kawna was not a regular miner, working the alluvial deposits along the lower end of the valley only on occasions of extraordinary opportunity or need, but he was skilled, deploying his experiences at panning into the present moment. And Kawna is exceptionally intelligent, knowing the land, remembering its past state as an intact rocky wall under which his present task once lay. As the layers peeled back over time, they revealed previously unseen textures of sediment—folds of sandy soil, mixed with glistening quartz and stone. On the day we stood together watching Kawna, we all attended to their unfolding with an eye toward the past (changing states of the rock) and the future (a potential windfall should gold be found). Kawna referenced the mythical associations between Biangai and this place as he dipped his pan, filling it with the historical, poetic, and physical qualities of the land. The entirety of this moment was part of a temporal flow that was unfolding in front of us, or what the early twentieth-century philosopher Henri Bergson calls *la durée*. For Bergson, *la durée* (duration) is characterized as continuous and heterogeneous, a multiplicity of moments that is forever folding time into space (Bergson 1950, 1971).[1] If gold were found, this would be an important and expansive event, setting off a series of social and ecological responses (Vayda and Walters 1999).[2] Kawna carefully filled his dish with old sediments and rushing water—with traces of the past and the endless possibilities of the future. His experience with this specific landscape, and gold-rich landscapes like it, led him to examine it carefully. Earlier in the morning, he had recalled other gold finds revealed in similar circumstances. Such experiences interpenetrate one another, folding themselves with Biangai histories of place, myth, and changes in the land. Likewise, the ecological world is both a very real material presence and a substantial site of meaning, gathered together in indigenous understandings of sociality (e.g., Biersack 1999; Roberts et al. 2004; TallBear 2014).

Following Bergson, French philosopher Gilles Deleuze (1988) suggests that objects (the gold pan, the gold, Kawna's hooded fleece) do not exist *in* time and do not have *a* time, but exist *through* time. Through time gold is caught up in the flow of sediments and the flow of ideas. But more important is the creative potential that such spatiotemporal relationships have in Pacific dance (Ka'ili 2017; Tengan 2008), cloth (Addo 2013), architecture (Refiti 2017), and the arts (Māhina 2017). While distinct from a larger Polynesian theory of space and time, my Biangai friends and family find creativity in their social and economic relationships, which are ultimately mediated over time through material networks of place. In particular, gold becomes invested with meaning, emerging as a real physical presence in the alluvial miner's dish. It simultaneously exists within a mytho-historical time, where gold is formed from the body of the first man and in alluvial deposits awaiting discovery, as well as the temporalities of markets with their fluctuating prices. For Kawna and his onlookers, places too have such a becoming, existing through time, defined as multiplicities where appearances of the past are folded into the present.[3] This is because Biangai places are animated with agencies that are both separate from the human subject and part of our very being. The land has duration: a past-future assemblage composed of human/nonhuman relations that are folded into the land in what I have termed "appearances of the landscape" (Halvaksz 2008b).[4] Such appearances reference a landscape's aesthetic qualities, simultaneously as physical surfaces and ghostly places rich with histories of human/nonhuman practices and meanings. Appearances attract gardeners to and repel them from the garden, and small-scale miners to and from the alluvium. They define both Biangai subjective and spatial relations as *placepersons*—collapsing Western distinctions between self and an earthly other, making mining for precious metals a particularly transformative task. The unearthing of sedimentary layers as they plant crops and dish for yellow flakes is a very real archaeology of embedded knowledge, which not only connects Biangai with the depths of their past but also accesses resources that are significant for how their future will be made (see Ka'ili 2017). It is from within this conceptual frame that Biangai approach development, where different ecological relations can productively collide (cf. Kirsch 2014)—not as a project of neoliberal economics, but as an ongoing project of folding the spatial and temporal relations among persons, places, and things of the past into the novelty of the present. In fact, they often resisted my critical perspective on their prospects. While I had hopes of offering advice to families that I had come to know and care for about the pitfalls of working with corporations and other neoliberal institutions, they remained hopeful. They cared less about Western academic critiques of conservation and mineral development, and more about how they might reimagine the future. Such positions are not naive and should temper the critical gaze of the Western academy.

Biangai assemblages of time and space are not the only ones at work in their valley. Their practices of gardening, cash cropping, and ecotourism are intertwined with Western capital, biologists, miners, tourists, geologists, missionaries, and anthropologists, as well as Papua New Guineans from throughout the young nation. Places across the landscape, the flora and fauna that adorn it, as well as the layers of soil, the rivers that cut through them, and the minerals that are revealed are more than part of the Biangai landscape. They are also part of what it means to be Biangai and what I'm calling placepersons.[5] Like naturecultures (Haraway 2003), placepersons force us to examine the blurring boundaries of place and subjectivity as places and persons mutually care for one another. Biangai might refer to such relations with the land as *ngaibilak* (cared-for ground). How do Biangai and their interlocutors utilize and make meaning with the physical world in which we all live? The image of Kawna panning for gold offers us a moment of contact (e.g., Pratt 1992) that assembles together Biangai places and meanings with a Western-inspired resource practice, where nature and culture are simultaneously mined for the market and exist within the conceptual frame of a distinctly Biangai way of knowing and understanding.

This ethnography is about the making of places. Further, it is about the relations between a physical place and its biological inhabitants, between ideas and environment, and between the practices of resource extraction and resource conservation. Such relations are made through the activities of those who call it home, those who seek their fortunes or who want to help others seek theirs, and those things we commonly call nature: mountains, rivers, streams, klinkii and hoop pines, yams, coffee, jewel beetles, gold, birds, tree kangaroos, and cassowaries.

Places, persons, and things are mutually constituted, but the perspective from which we view their becoming is not uniform. The Bulolo Valley is a negotiated space, subjected to practices of neoliberal capital, the logic and classification of biologists, geologists, and the like, the sociological imagination of ethnographers, and most importantly, Biangai worldviews. The physical space and its inhabitants have no single author, as suggested by Kawna's efforts to find gold, a mineral that is valued in Western centers of calculation, amid the collapsing site of a great ancestor's fall. Distinctions between Biangai relations to place and neoliberal practices of the same are comparable to twentieth-century philosopher and literary critic Mikhail Bakhtin's (1981) distinction between the dialogical novel and the rooted epic (Halvaksz 2013a). Like the many stories that Biangai tell about land and their placement, the novel affords many different conversations, while the epic persists as singular, dominating voice of the past. Whereas Biangai places are multiplicities, law and legal ownership bind neoliberal places into an epic form. My own efforts here seek to contribute without limiting.

The physical space is also an author, a source of knowledge, a mediator between epistemological systems, and an object engaged for political and economic endeavors. The landscape inscribes its novel contours into the bodies, social relations, livelihoods, and imaginations of those who transverse and burrow in its surfaces. It creates, folds, and reveals historical seams of precious minerals. Its rivers nourish the production of coffee and subsistence crops, and devastate them with rushing waters and floods. Prospectors are drawn down into its layers, and biologists scour its surfaces for rare species, unique ecological relationships, and animal behaviors. Through both conservation and prospecting, it is a mined space.

More than ninety years of gold mining, conservation, and biological research wrought transformations in the landscape. The Biangai struggle to understand why, in spite of the golden wealth that comes out of the ground around them and the sustained interest in their flora and fauna, they continue to rely on subsistence horticulture and coffee, a rather temperamental cash crop. And they continue to look positively toward the future. Three themes connect these "entangled landscapes" (Moore 2005): assemblages of natureculture, theories of place, and neoliberal development. These frameworks can help us understand Biangai senses of place and their woeful placement within a global system of inequality. That they remain hopeful, ambitious, and even convinced of their own rightful place at the center of global flows of power and wealth in spite of great structural inequalities is inspirational. I hope that my rendering of their lives can capture their spirit and ambition, as well as provide a critical perspective on current practices of neoliberal conservation and resource extraction.

ASSEMBLING NATURECULTURE

Nature and culture are problematic categories within anthropology. Taken together, they have been viewed as metaphorical and real oppositions defining place, gender, and in the worst case, race. But in contrast to structuralist accounts that see nature as feminine and culture and masculine (e.g., Ortner 1974), nature and culture also reflect one another, compose one another, and at times become indistinguishable as natureculture. Many have questioned the validity of the categories outside of Western discourse and society, calling for greater attention to indigenous ontologies and epistemologies.[6] Recent ethnographic work has pushed the boundaries of the debate further to consider the impossibility of truly understanding the worldview of another and to write ethnography so as to let radical alterities coexist (e.g., the ontological turn in anthropology). As Eduardo Viveiros de Castro argues, "Anthropology compares *so as to translate*" and a "good translation is one that allows the alien concept to deform and subvert the translator's toolbox so that the intentions of the original language can be expressed within the new one" (2004, 5, emphasis original). The emphasis in this work is on the subversion of the ethnographer's modes of translation, such that indigenous perspectives prevail.

However, the ontological turn has been justly critiqued for the failure of advocates to consider both feminist and indigenous scholarship that long ago laid claim to this intervention as a project of decolonization.[7] In indigenous approaches, place-thinking subverts considerations of Western ontology, our intellectual genealogies are examined, and the possibility of a shared human and other-than-human world is envisioned. Writing about indigenous "place-thought" in North America, Vanessa Watts argues, "Place-Thought is the non-distinctive space where place and thought were never separated because they never could or can be separated. Place-Thought is based upon the premise that land is alive and thinking and that humans and non-humans derive agency through the extensions of these thoughts" (2013, 21).

The objective is to figure out "how to share goals and desires while staying engaged in critical conversation and producing new knowledge and insights" (TallBear 2014, 1). Such perspectives are not taken to create boundaries between different ontological positions. Instead, a critical perspective is required to engage and critique structures of power and authority, and to improve the lives of those involved. While I take seriously the mutually constituted relationship between nature and culture implicated in the ontological turn (Graeber 2015), the goal here is "to show that in certain ways, at least, such alterity was not quite as radical as we thought, and we can put those apparently exotic concepts to work to reexamine our own everyday assumptions and to say something new about human beings in general" (6; see also West 2016, 112). By exploring Biangai ideas of nature and culture—or at least how their worldview might inform our concepts—we may be able to articulate mutually inclusive future ecologies.

Naturecultures link human and nonhuman worlds. But posthumanist scholarship argues that it is not simply a matter of creating these links. They already exist. The problem has been that we don't recognize them. Within Western discourse and practice, such positioning has required an active reimagining by demanding the creation of a new constitution (Latour 1993), rethinking Western relationships with nature (P. Robbins 2007) and our animal companions (Haraway 2008), and proposing radical transformation of economic, political, and social life (Kirksey 2012, 2015). By contrast, Pacific epistemologies have long emphasized these relationships, noting the importance of the physical environment in conceptualizing how communities know and experience the world.[8] Biologist Mere Roberts and her colleagues (2004), for example, argue that Maori genealogies, or *whakapapa* (literally, "to place in layers"), not only reflect human ancestry but also narrate relations among persons and places through the many species that coinhabit the land. They note, "Insects and humans, fish and ferns, stars and stones all descend from the spiritual realm of the atua [gods], and thus all possess spiritual qualities in addition to their own unique material attributes" (4). Likewise, Biangai relations to place evoke a shared territoriality with the flora and fauna, linking nature and culture in ways anticipated by posthumanist ideals and reflective of interventions by indigenous scholars in how we live in the world.

Within Papua New Guinea the opposition of nature and culture has certainly been critically engaged (e.g., M. Strathern 1980; R. Wagner 1975). Arguing against a Western essentialist view of nature and culture, anthropologist Marilyn Strathern shows that for Hagen society specifically, and Melanesia more generally, domestic and wild are not easily separated into a male culture and a female nature. Instead, domestic and wild are generative: "Neither male nor female can possibly stand for 'humanity' as against 'nature' because the distinction between them is used to evaluate areas in which human action is creative and individuating" (1980, 219). As Strathern notes more broadly, the relationship between persons and materiality is quite complex.

Biangai too conceptualize the world in a way that separates wild from domestic. As with other things, wild and domestic are discussed as processes of becoming one or the other. Through labor the wild is domesticated, while left untouched the domestic can be made wild. The forest, Yalamu told me, is the source of cool breezes but also wild game, and potentially a garden. It is dangerous and yet knowingly looks after its own by revealing places for gardening, hunting, coffee, and so on. It is a matter of listening to and knowing the land, he explained. As Yalamu suggests, nature and culture are never entirely lost but are always potentially becoming a source of the other. Hope for the future rests on these messy entanglements.

Disentangled in Western thought, nature and culture are both resources to be mined. Nature is contracted, extracted, and sold as mineral wealth, biological specimens, timber, cash crops, and so on. In its many forms, nature is made mobile to support academic careers and profits. At the same time, indigenous peoples deal with a sense of loss in cultural and linguistic identity as practices and beliefs are subjected to the powerful forces of capital, missions, and government. Anthropologist Anna Tsing (2015) calls such precarious spaces third nature, exploring what manages to live in spite of capitalism. In her ethnography of Matsutake world-making, she evokes the patchy ruins where relations with place are uncertain, changing, and challenging. Such ecological patches give rise to "emergent ecologies" (Kirksey 2015). This suggests possibilities for naturecultures that exist in the detritus and ruins of neoliberal capital. But the concept of placepersons offers a different vision of ecological emergence. Precarity and patchy world-making might be a condition of urban life in Papua New Guinea, but for Biangai, connections to place mitigate such uncertain worlds.

PACIFIC PLACES

At the heart of natureculture discourse is how human beings encounter, engage, create, reflect, deny, and even speak of the world. Too often we have limited ourselves to vision while at the same time celebrating the power of language (Serres 2009). How we see and speak of nature takes precedence over other ways that we might sense it. In place of such narrow sources of insight, the skin, through which the body encounters the world, is a permeable boundary that importantly guides all of our senses, such that nature and culture, places and persons, bleed into one another. The world enters our flesh, but for many in Papua New Guinea, flesh likewise covers the world.

Composed of the shed skin of a python ancestor (Biersack 1999) or taking the form of totemic animals (e.g., Bulmer 1967), important trees (e.g., O'Hanlon and Frankland 2003), or stones-ancestors (e.g., Kahn 1990), Papua New Guinean places and human bodies seep into each other. They are "inseparable" (J. Weiner 2004, 3). For the Paeila, Biersack notes that the totemic snake-ancestor, Taiyundika, locates and symbolizes organic life: "As any Paiela can and does attest, snakes molt, replacing skin with skin, parthenogetically reproducing themselves out of themselves, body out of body and flesh out of flesh" (1999, 74).[9] Shedding its skin, the python renews its connections with organic life, sensing the bodily presence and sacrifice. Through its skin-earth it senses the world. For Biangai, the paths, gardens, and hunting grounds are occupied spaces rich with spirit-ancestors and human actions. Lands are referred to as *ngaibilak*, which literally means "cared-for ground." But in this sense the caring is mutual, as garden and hunting lands care for those who grow from their relationships. The landscape is an extension of their bodies, reading the actions of Biangai as they work their gardens, follow paths in the hunt, journey between communities, consent to mining or logging, and invite researchers to their lands. Placepersons are made through these relationships as the world enters into their flesh, and at death, the flesh becomes part of the world. In pre-mission burial practices, the skull and long bones of Biangai bodies marked these relationships as they were placed in significant locations throughout the land, mingling person and place in death as they were mingled in life. These were also markers of deeply held spatial relationships among the living and their ever-present ancestors.

More broadly, places on land and sea are central to an understanding of how Pacific peoples engage with the world (Hau'ofa 1993). The sea, Hau'ofa notes, opens up Oceanic places, expanding the worlds of island communities and subverting notions of isolation and the diminishment of their economic potential. By contrast, inland communities conceptualize places on land, defined by a sense of rootedness (O'Hanlon and Frankland 2003). The two are intermingled in Oceanic thought and scholarship, as the "natural landscape" is marked with "maps of movements, pauses, and more movements" (Hau'ofa 2008, 73). While places on land and sea are both significant in a broader Oceanic identity, our focus here is on the interior lands of Papua New Guinea. However, the relationships to time and space retain certain similarities.

In outlining an approach to theology, Fijian scholar Ilaitia Tuwere (2002) argued for the centrality of *vanua* as a guiding principle. *Vanua* literally means "land" but symbolically encompasses livelihood, the relationship of time and event, connections to the ancestors, and custom (33, 35–41). Tuwere argues that the land is all-encompassing and central to Fijian spiritual life. Nabobo-Baba pushes this concept farther, grounding her research on knowledge practices in relation to the land. She argues, "Vanua research supports and affirms existing protocols of relationships, ceremony and knowledge acquisition. It ensures that the research benefits the vanua and that the love, support and resources given by the people are appropriately reciprocated" (2006, 25). By literally *grounding* research in the relations among persons and places, she emphasizes a shift in ways of knowing, from epistemological questions of how we know to how things are related across space and time.[10]

Other indigenous Oceanic scholars have likewise pointed to spatial and temporal relationships as guides to Oceanic understandings of the world. Notably, *tā-vā* scholarship and research advocated for by the Tongan philosopher Māhina (2010) have called for Oceanic researchers to examine time-space relationships as indigenous categories. Like others (see Ka'ili 2017; Ka'ili, Māhina, and Addo 2017), he sees indigenous scholarship as focused on the plural, holistic, and collective, as well as temporally circular experiences of time and space (*tā* and *vā*) common to Oceanic thought. Time, in this sense, is a rhythm and a motion or "a beating of space" (Māhina, quoted in Ka'ili 2017, 38)—like an artist, or dance, where spatiotemporal worlds intermingle through the beating of a drum or the artist's brush. For Māhina, space is highly social and binds relationships into exchanges over time. Furthermore, Ka'ili calls attention to the Tongan concept of *tauhi vā* as a spatiotemporal practice "to create and preserve harmonious and beautiful social spaces within all groups and all sectors of society" (2017, 33). He argues that different spatial practices of individuals are woven or folded together to create beauty in the world. The *tauhi vā* is more than space-time, encompassing the rhythm of relationships and exchanges among groups, which is both creative and productive.

These concepts have equally important implications for how we think about "development." In Samoa, *vā* likewise connotes social space, "the space between, the betweenness, not empty space, but space that relates" (Wendt 1999, 402), not a gap or opening. For Wendt, Samoan *vā* shifts us away from Western ideas of development by emphasizing indigenous conceptualization of place as an important way to think about economic practices attending to different metaphors for economy (see also Gudeman 2016). Space does not need to be filled by development, because it is already made meaningful, and indeed beautiful, by the relationships it evokes over time. As Lilomaiava-Doktor has argued, reading development through the principles of *vā* allows us to "focus on support and caring for relationships rather than the pursuit of wealth for wealth's sake" (2009, 20).

Biangai *ngaibilak* practices are likewise generative of historically and socially important relationships by folding space and time into a unique aesthetic or appearance (see chapter 6). While it is important to note that as non-Austronesian speakers, Biangai lack a concise expression for *tā-vā*, they tend to the land and to each other in similar ways: by folding time and space together in productive relationships in the garden (chapter 3), among kin (chapter 2), as well as in relationship to development (chapters 4, 5, and 6). Time and space are "gathered together" with "multiple pleats" (Serres and Latour 1995, 60). It is also central to the processes of panning for gold. While Papua New Guineans are linguistically distinct, they do share common ways of relating to place.

Papua New Guinean academics have likewise emphasized place and places in their scholarship. The poet and scholar Steven Winduo describes the significance of rivers in his book of poetry, *Hembemba* (2000), as rivers relate to the places where he has lived. In an interview he stated, "*Hembemba* in the Nagum Boiken language means 'river of the forest.' When Nagum Boiken use the expression *hembemba yawi nangu*, they mean one's place, home village and all the things that sustain one's life" (Wood and Winduo 2006, 85). The river of the forest is inseparable from the environment and the community, and made deeply meaningful in Winduo's poetry. As Bougainvillian author Regis Stella argues, "Land and place cannot be separated," because "land/place is the center of indigenous constructions of social histories, of personal and interpersonal experiences. So in addition to ownership of place as such, for Papua New Guineans, landownership connotes identity, belonging, and social existence" (2007, 29). While linguistic and cultural diversity preclude naming a single concept like *va*, place is widely seen as generative of socioeconomic relationships in Papua New Guinea. In the lingua franca Tok Pisin, Papua New Guineans typically refer to *papagraun* or *mamagraun* to denote these shared ideas. While "fatherland" and "motherland" might be literal translations, the sense of nationalism associated with these constructs in the West is misleading. Instead, it is more about the relationships between kinship groups and land. Again, caring for places is lexically evoked. Places tie social relationships together through shared knowledge, labor, and lineage, as a source of communal agency and creativity (Leach 2003; Thornton 2015). They are most certainly produced (Lefebvre 1992), but also expressions of intimacy and love (Bachelard 1964, xxxv).

Importantly, it is the productive nature of the landscape that comes to demarcate these relationships to place, creating parcels that circulate.

Geographer Yi-Fu Tuan tells us that "place is a type of object" and that "places and objects define space, giving it a geometric personality" (1977, 17). The circulation of places and objects not only gives space a geometric personality; it also creates and maintains social relations over time as garden lands and hunting rights move through and with families. This is especially important with respect to garden produce, pigs, and other culturally defined valuables (such as those in the *kula* ring in Trobriand Islands described in Nancy Munn 1986), which are contributed to exchanges. Agency in such contexts has been described as "the strategic detachment, decomposition or deconception of part of oneself and [the part's] attachment or composition as part of another person" in an endless network that incorporates kin and nonkin, human and nonhuman entities (Mosko 2001, 260, following M. Strathern 1990). Objects of exchange, and goods in general, are part of oneself. But following Pacific scholars and Biangai sensibilities, they are also parts of one's place, and places are central nodes in this wider network, which is both creative and procreative. Persons are socialized in the same interactions that socialize tangible (shell valuables, trees, gardens, forest paths, pigs, etc.) and intangible (like songs, magic, stories, etc.) entities. It is a process of detaching and attaching parts of subjects to form *objects and places* as well as detaching parts of *objects and places* to form subjects. As detachable parts of persons, objects and places become alive with a kind of agency of their own (see also Watts 2013). Persons become places, and places become persons.[11]

An ethnography of such practices must attend to the "grounding" of human life. As James Weiner notes for Papua New Guinea, "The land speaks . . . but only through the human voice and hand in their appropriative manipulability, from which we are unable to dissociate the contribution that poetry and myth make with regard to their own historical and deictic functionality" (1998, 137). Papua New Guinean persons, their voices and hands, are grounded both through interactions with the earth and their aesthetic imagination. Grounding engages practices of work associated with living on the land.

Subsistence gardens illustrate this relationship as the forests are repeatedly cut in cycles that reflect the availability of land, quality of soil, and needs of the population. But most importantly, they are produced and then reproduced by networks of kin who exchange and share specific parcels. As with other elements of the landscape, gardens are objects that are named and "circulated," embedded with the labor of both the living and the dead. Only those who have rights to such lands can be successful on them. However, this is not because they are "property" but because they are a real part of the social lives of the community (Māhina 2010). Thus, gardens become inalienable objects—hybrids of nature and culture—that are produced, reproduced, and exchanged by women and men as they go about their lives. By digging into the soil once turned by one's ancestors, one folds time, space, and genealogy together into a present relationship, while creating the potentiality for future sociality. Working the land makes persons both viable producers of exchange items and members of a broader group with connections to a specific resource (Fajans 1998). Gardens are but one example, as the locations of trees, streams, houses, roads, paths, and many other places are richly connected with ancestral and personal histories. As a result of these histories, many Papua New Guineans in general and Biangai in particular imagine the landscape itself as active with relational agencies that take on environmental and ancestral forms.

For example, the Biangai speak of coffee plantations being developed in Wau township with a great deal of skepticism. They view the land as fertile yet unproductive. This land was alienated by the Australian government, and its contemporary owners and managers do not hold ancestral claims. Biangai conclude that those from outside cannot successfully work this land because the relationship between people has been confused. By viewing such places as hybrids of human and nonhuman networks, we can begin to understand how "cutting the network" (M. Strathern 1996) could lead to the land's infertility. When the relationships between Biangai and places are severed through colonial alienation, logging, mining, or even conservation, the ability of the land to be productive is affected (e.g., Taussig 1980).

NEOLIBERAL NETWORKS OF DEVELOPMENT

Neoliberalism might simply describe the most recent moment in capitalism, as the market is increasingly freed from government regulations and commitments to the nation. In place of pre-WWII state involvement in labor, markets, finance, and trade, or what cultural geographer David Harvey (2005) calls embedded liberalism, post-WWII markets de-emphasized the moral economy valued by a more interventionist state. While in the West this has facilitated great concentrations of wealth within corporations, global financial institutions have pushed for increasing deregulation from countries that primarily supply raw materials, such as Papua New Guinea. Further, as Marx might appreciate, neoliberal capital finds in such locations a more susceptible, if not more vulnerable, body of laborers. As a result, a regulatory environment that favors international corporate interests over both state and community characterizes neoliberalism.

Importantly for my argument, Michel Foucault (2008) suggests how laborers working for development projects such as a gold mine or a conservation area can be understood to embody neoliberal practices. Through specific technologies of power, they are fashioned into neoliberal subjectivities to facilitate competitive markets driven by cost-benefit analysis and self-interest. From the viewpoint of neoliberal development, capital requires the "creative destruction" of "prior institutional frameworks" including "divisions of labor, social relations, welfare provisions, technological mixes, ways of life, attachments to the land, [and] habits of the heart" (Harvey 2005, 3). In regard to the mining camps and work sites, this destruction is apparent and highly regulated by international regulatory mining frameworks and national laws (Jacka 2015; Kirsch 2006, 2014). And a growing body of literature highlights the transformative power of market-driven conservation to shape social life.[12] In this volume, I examine the degree to which conservation and mine workers embody neoliberal ideals that could be transformative for nearby village economies and social life. If neoliberalization includes opposition to collective projects and the celebration of individuality, flexibility, and self-realization,[13] to what extent do mining royalties, ecotourism labor, compensation, village education, and employment create neoliberal subjects? What form do these subjectivities take in a Melanesian context? How do these subjectivities shape daily life in nearby residential communities, from which mining employees are drawn? Are subjects drawn away from the commitments of community life? Or do they resist these technologies of power?

Neoliberalism is not a thing that acts independently. Instead, it is what philosopher Bruno Latour (2005) might call a panorama, "an illusion of totality" to which we might subscribe. But in contrast to Latour, it is not simply a vision of the world that one taps into, or downloads, to use his metaphor. If we are to truly understand neoliberalism, we cannot take it for a simple structural abstraction. To understand neoliberalism we must trace its conduits as subjects, objects, and places are *neoliberalized*. Viewing neoliberalism as a network does help us see its many paths. Some paths are freely accepted and even desired in places like Papua New Guinea; some conduits, however, come with great force (see Harvey 2005; West 2012, 2016). As Biangai struggle with neoliberal regimes of development and government, they struggle to control the flow of the latter even as they seek out the former.

Neoliberal practices, then, are transformative of landscapes but also the body (Foucault 2008). Neoliberalism carves the skin, dividing the senses, disrupting flow and movement. The body is reduced to parts that can be made into partible subjectivities. Workers who lose fingers, hands, or the functionality of muscles, who are worn and injured from laboring at the mine, are compensated for their loss. These parts of oneself do not circulate in any meaningful way but are monetized within the context of neoliberal capital and compensation. The earth, too, is like the body, scarred and explored by neoliberalism's fascination with nature. For Biangai, transformation of the ground is felt poetically (Halvaksz 2003). The body, like the land, is segmented by the regimes of work and rest; it is revalued. Foods taste different (Halvaksz 2013b); songs are rewritten for a different set of ears (Halvaksz 2003). Places and persons are transformed by neoliberalism's power even as participants remain awed by its advertised potential to improve their lives (Halvaksz 2013a). It is a new refrain, a territorial assemblage of great power (Deleuze and Guattari 1987), whose rhythms are measured in time and space. And yet, Biangai persist, attracted by the precarious promises of capital even as they find respite in the land.

This book is meant to contribute to our understanding of how neoliberal economic practices shape place-based economies and identities by focusing on local responses along the frontiers of resource extraction (Tsing 2005). For example, in the context of conservation, neoliberal policies can "work as biopower to construct and regulate life and lives in significant ways" (Büscher et al. 2012, 5). That resource extraction likewise reconstructs and regulates the lives of workers on site is well documented,[14] but how does working for the mine transform work and social relations in the communities that supply labor and benefit financially? What does conservation mean to those who supply field guides and look after the area? And what specifically might this mean for placepersons? Neoliberalism differs from an older form of market liberalism in that it seeks to expand market rationality, creating other spaces for rational market-like decisions (Fletcher 2010). Neoliberalism is a process calling for research that attends to the neoliberalization of spaces around capital (Peck and Tickell 2002). Neoliberalism flows from one site of production to another, through the objects and subjects that are fashioned by its becoming—from centers of capital to mining sites and conservation areas, and from these bounded spaces to the community. This raises questions about how it flows from one place to another: if mining and conservation are productive of new subjectivities, how do these subjectivities contribute to the creative destruction and reconstitution of village life? Do mine worker families embody principles of neoliberalism in their decisions about gardening practices and village labor? In other words, how does neoliberalization shape social

FIGURE I.1 Performers from Elauru's Lutheran Church youth group welcome guests to a weekend service organized by Biangai women from throughout the valley, 2011.

reproduction, especially when those socialities include more-than-human relationships?

Neoliberalization of village life is evident in changes in agricultural practices, where community-oriented livelihood strategies, relationships, and landscapes (Harvey 2005) are still salient. Attention to agricultural change is not new. An older literature was attentive to agricultural intensification with the introduction of market economies, new technical inputs, the commodification of crops, and was often grounded in debates around formalist, substantivist, Marxist, and ecological approaches.[15] Much of the literature on globalization, neoliberal governance, and agriculture calls attention to state policy, and regional and global governance, as well as increasing concern with food security/sovereignty, and rightfully so.[16] However, attention to the relationship between the different scales at which agricultural policies are created and enacted and historical attention to micropractices of subsistence gardening reveal the linkages that make such changes possible.[17] Few studies consider the transformation of agricultural practices and subjectivities in

FIGURE I.2 Winima community members gather for a Saturday morning market where produce and other items are sold, 2011. The health clinic and schoolhouse built by the mine are prominent features at the center of the community.

the context of industrial resource extraction.[18] Even with increased attention to practices of corporate social responsibility and community-company relations,[19] we know little about community agricultural practices as spatial, ecological, and social products in the ambit of a mine. Biangai narratives and practices reveal a certain resiliency in the face of such transformations, but even this is not perfect. The specific processes of agricultural change, which are largely minimized in studies of mining, reveal the wider impact that work outside the community has on work inside the community.[20]

Agricultural change is motivated by social relations as well as increased subsistence and commodity demands (Brookfield 1972, 1984). The social role of food production is as important as materialist concerns with simple agricultural intensification and subsistence production (e.g., Boserup 1965; Watson 1967). Kinship, land rights, labor commitments, as well as ideational factors, must be considered to understand changes in production. Cultural geographer Harold Brookfield called this *social production*. While much of contemporary research focuses on broad changes in agricultural practices, we must attend to how neoliberalism affects the overall social production of a society.

Thus, if agriculture is increasingly about producing cash crops and not about the reproduction of social relationships through planting yams and other subsistence foods, or if a neoliberal work ethic is transforming agricultural production, how does this shape the relationship between place and person? What becomes of community-based subsistence gardens and smallholder agriculture in such a context? How is the social *re*production of the community accomplished? It is this dilemma of mining, conservation, and changing local and regional economies that this book addresses.

Mining and conservation offer a unique position from which to examine the intersections of natureculture, indigenous ideas of place and neoliberalism. Especially as each of these conceptual domains is at play in debates among Biangai about how they imagine their future and retain hope that their lives will get better. It is from a sense of place, grounded in productive relationships between nature and culture, that Biangai resist global forces and remain hopeful. Place connects Biangai to one another, making the recruitment of places into mining and conservation both troubling and full of potential.

THE UPPER BULOLO VALLEY

Shaped by indigenous, scientific, and colonial histories, the Upper Bulolo River Valley has a prominent place in the story of the nation. After gold was discovered by prospectors in the 1920s, the valley became the site of one of the largest gold rushes in the Southern Hemisphere (Waterhouse 2010). As a result, it typifies the sort of place that early anthropology ignored—a place intimately connected to global mining, missionaries, and government agencies. Biangai communities were largely on the margins of these developments, selling some produce, learning the practices of the Lutheran Church, and occasionally working with miners from all over the world. But these transformations offered communities enormous contrasts in material worlds. With many of the conveniences of Western life, the town of Wau even publicly competed with coastal cities to be the capital of the territory (*Canberra Times* 1938; *Papuan Courier* 1938). Of course, lacking a port was a major deficit. World War II intensified connections between the valley and the world but also represented a shift in the capital investment in the region, as much of the mining infrastructure was destroyed in Japanese raids (Bradley 2008; Waterhouse 2010). Coffee, agriculture, and logging were imperfect replacements but maintained an outward gaze toward the world. Wau's airport once handled as much cargo as all of Australia combined, but its sloped grass field was incompatible with the advances in technology of newer planes. With air transport replaced by barely maintained roads, the communities in the valley were not so much isolated as their connections to nation were weakening. By independence, colonial control of political and economic life in the valley was replaced with local agencies.

Before the first miners arrived, Biangai lived in seven distinct communities organized with individual homes along a central path. The typical community was described as being surrounded by staked fencing held together with tightly woven vines and bark ropes. Men's cult houses were located at the ends of the row of houses. Villages were organized around these local men's houses, where young men resided with senior males. Women's houses lined the main path through the community. From stories and historical accounts, Biangai viewed this as a time of warfare with neighboring Watut-speaking communities. But warfare among Biangai communities was also common (Burton 1996b). The men's houses were organized around a singular mythical figure (whose name I cannot reveal), and a central men's house was maintained in his honor. No living Biangai had participated in the initiations associated with the precolonial men's houses, and much of their experience with them comes from stories told by their parents. Today, villages are large, unbounded territories, expanding and contracting with shifts in population. Families reside near one another, and the practices of the men's house are lost. Mining, logging, coffee production, and the occasional tourist continue to forge the connections between Biangai places and the rest of the world, but perhaps with less intensity than witnessed during the early stages of colonialism.

Linguistically, Biangai belongs to the Kunimaipan family within the broader Papua phylum of non-Austronesian languages (Dutton 1976; McElhanon 1984). Biangai origin stories and linguistic cognates suggest close relations with Biaru and Waria speakers to the east (Chinnery 1928). Angan speakers occupy areas to the southwest,[21] and Austronesian groups separated by the Kuper Range occupy coastal areas to the north (see J. Wagner 2002, 2007). By recent estimates there are at least 6,223 speakers of Biangai (National Census Office 2014), divided into three dialects (see Dubert and Dubert 1973). Elauru, Winima, Werewere, and Kwembu speak one dialect, Wandumi and Kaisenik another, and Biawen, the central village, shares aspects of each (Dubert and Dubert 1973, 5).

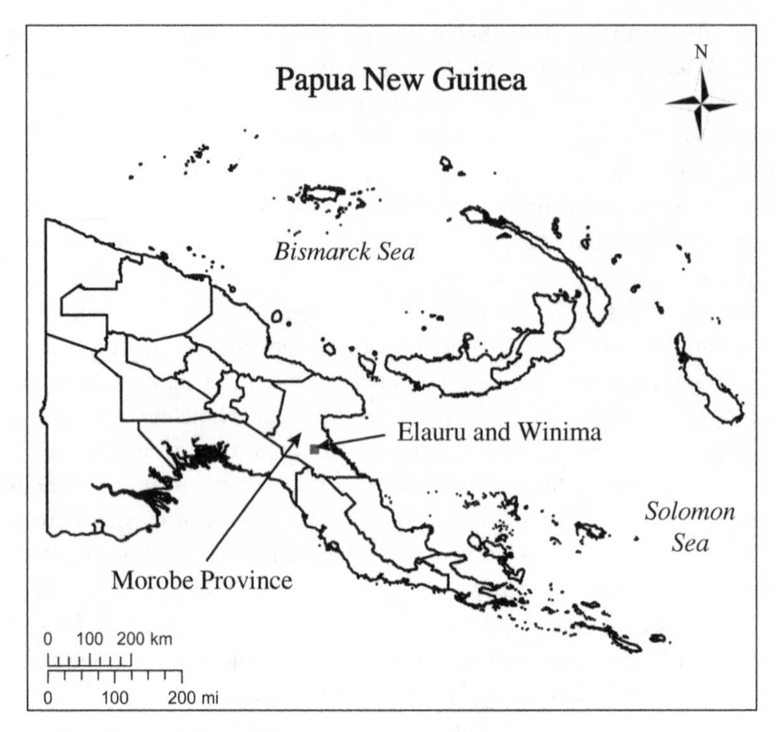

MAP I.1 Morobe Province and Papua New Guinea

Fieldwork Sites

While I am familiar with all of these communities, research was conducted in the neighboring communities of Winima and Elauru at the far end of the valley (map I.2). My residence alternated on a weekly basis as I followed community members' divergent participations in distinct development projects. Beginning in 1990, Elauru participated in a now-defunct effort to organize a conservation area in the Kuper Range. The Wau Ecology Institute (WEI) was responsible for organizing this effort in collaboration with village leaders. The oldest conservation NGO in Papua New Guinea, WEI was founded by the Bishop Museum (Hawai'i) as a field station in the 1960s before being turned over to Papua New Guinean management after independence. During my initial research (1998, 2000–2002), the conservation area was alive with activity. By the time I returned in 2005, it had ended in dramatic fashion. Subsequent trips in 2011, 2014, 2015, and 2016 found conservation an ongoing part of the local discourse about development, but with few opportunities to realize its perceived potential.

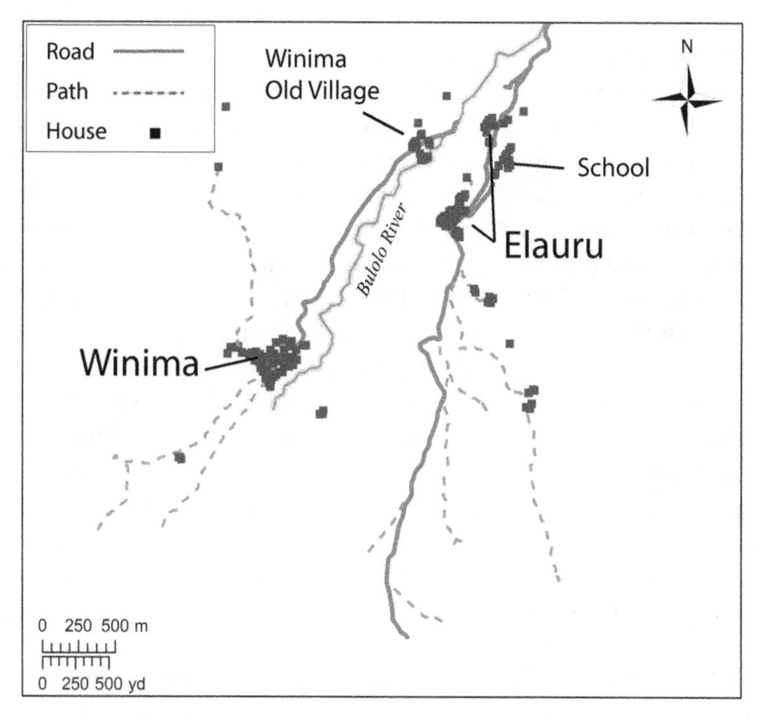

MAP I.2 Elauru and Winima

By contrast, Winima shares surface rights to Hidden Valley Gold Mine, operated by Morobe Goldfields on behalf of Harmony Gold (South Africa). This is the most recent manifestation of corporate operators, as the mineral rights have changed hands numerous times since formal exploration started in 1984. Under the current employment arrangements, Winima villagers have work opportunities with Morobe Goldfields or in supporting industries. Elauru villagers rely on the occasional ecotourists and limited employment with the mining industry. Both are active in the production of coffee and cash crops for local markets. While residents in both Winima and Elauru lack the prospects for artisanal mining found among villages closer to Wau, they have had opportunities to work the leaseholds of kin and on the land of Biawen, Wandumi, and Kwembu (Biangai communities closer to active gold-mining areas). And many older Biangai worked in their youth alongside whites who once dominated small-scale production.

Both communities maintain subsistence yam and sweet potato gardens, as well as participate in the production of coffee on land controlled by men and women through cognatic kinship practices. While most of the food produced in gardens is for household consumption, a small surplus does enter into local markets or is exchanged among kin. Yams are an exception as they are rarely sold. Instead they feed into special events most often associated with funerary arrangements. First introduced into the communities in 1955, coffee remains a steady source of income during the harvest season, often drawing urban Biangai back to their home communities to prune trees, clean the gardens, and pick the ripening coffee cherries. Those who don't return relinquish control of their groves to family or let the trees grow wild.

The villages are located less than five kilometers apart, with multiple paths and a single road connecting them. Winima is spread along part of the Upper Bulolo River, in two sites: an older village closer to the main road that has been inhabited since World War II, and a newer location farther up the valley dating to 2000. Both Winima locations are adjacent to the Bulolo River, placing the community in the middle of their primary coffee gardens and short distances from their main gardening areas. Owing to historical flooding, the topsoil of the valley is uniquely rich (about a meter deep in the main village), allowing for robust gardens and highly productive coffee trees.

Uniquely, Winima is a linguistically diverse community, as families have long sponsored Watut settlers on their land. Based along the Watut River Valley, the Watut were traditional enemies of Biangai prior to colonization.[22] While Upper Watut communities are beneficiaries of the mine, settlers in Winima are from other communities within the wider Angan linguistic area. Between twenty and twenty-five Watut households were established in Kudjuru, in a neighboring valley on Winima land. Separated by a high mountain ridge (a full day's walk from the village), they have little impact on daily life in the community, though they do vote in local elections. Within the main villages, Biangai likewise sponsor a number of Watut families (eleven households in 2016, or one-quarter of the active households in the community). Initially connected to Biangai families through church, these families look after absentee Biangai lands and work alongside sponsors in their coffee gardens. While a few individuals have married into the community, the majority are independent households. In exchange for this labor, they have shelter and a place to plant their own food, and increasingly their own coffee trees. Their position remains precarious, as Biangai can easily evict them. While marriage outside the community is increasingly common, Elauru had no practice of sponsorship.

Elauru is likewise divided between a "new camp" and an "old place," both of which are built on the ridge that the Wau-Waria road follows through the

Kuper Range. The older and larger village has been occupied since the 1950s and hosts the community school shared with Werewere and Winima villages. The "new camp" originated in religious distinctions between Pentecostals and Lutherans. Members of the Pentecostal Church wished to live free of tobacco, alcohol, and betelnut, and established the smaller hamlet. Today, this "new camp" is sparsely occupied, as most have migrated to town or back to the older village. Those with more distant gardens also established a third, smaller hamlet called Kambili along Naiyaka creek. The landscape around the village is dominated by steep slopes of kunai grasslands with small tributaries of the Naiyaka and Bulolo lined with shade-grown coffee. The soil is more typical of tropical soils in the region, with a thin layer of topsoil and a steep slope that facilitates erosion and landslides.

Ethnographic research near the Wau-Bulolo Valley was first conducted by E. W. P. Chinnery (1928; see also S. Chinnery 1998). His only observation regarding the Biangai was to note that they were of the "same type" as the linguistically similar populations along the Waria and Biaru Rivers. Based in Biawen, the anthropologists Jean-Claude Martin and Françoise-Romaine Ouellette conducted ethnographic research in Biawen village in the late 1970s on the transformation of both agricultural practices and gender relations in association with coffee production in the Biangai villages surrounding the town of Wau (Martin and Ouellette 1981). They utilized French structural Marxism to highlight the articulation of kin-based and capitalist relations of production and their effects on coffee-growing communities (Martin 1992; Ouellette 1987). While coffee continues to be pervasive, the processes that now affect Biangai speakers were not yet apparent during their fieldwork, and neither continues to do research in the region (personal communication, November 2011). Ngawae Mitio, a Biangai man from Werewere, completed his honors thesis in anthropology in 1981 on logging negotiations but continued to work in the area as a researcher at the Wau Ecology Institute, and eventually with Hidden Valley Gold Mine as part of the social monitoring and community relations team. Today, he teaches at the University of Technology in Lae. More recently, Daniele Moretti (2006) completed his dissertation on small-scale miners in Wau, largely focused on the migrant communities from the Watut areas. Finally, as part of the environmental impact assessment for Hidden Valley Gold Mine, John Burton conducted a detailed study of Biangai and Watut communities. His extensive research has provided valuable insight into the relations between corporate mining and landowner communities. While some data remains the property of the company, his reports have been made available (see Burton 1996a, 1996b, 2000a, 2000b, 2013).

Biangai claim that prior to colonial contact, they were the sole occupants of the Upper Bulolo Valley, encountering only the occasional hunting party from neighboring areas. They traded for shells with coastal communities to the north (J. Wagner 2007) and maintained kinship and exchange relationships with Biaru communities to the east. A state of warfare with neighboring Angan groups preceded colonial contact (see also Blackwood 1978; Martin 1992). While these precolonial times are remembered as a time of Biangai independence, Biangai mytho-poetic tales, at least in hindsight, favorably predicted the coming colonization. They foretold of flying houses, white skins, and guns, as well as new wealth and rice. Only through experience did Biangai learn that gold could be the source of such transformations. But more importantly, their stories predicted a moral transformation, where good behavior would replace conflict and cannibalism. While associated with the emergence of Lutheran missionaries and the civil order imposed by the colonial state, Biangai also reflectively associate such transformations with development. And along the Upper Bulolo development is derived from gold.

CHAPTER ONE

Mining Nature

By 1926, Biangai were at war with the many gold prospectors who had settled into their valley. At first, they kept some distance, and then increasingly traded produce for the curious items that were transported onto the goldfields by the many miners and carriers who made the trip up from the coast. The Australians and Europeans were few in number when compared to their carriers, who hailed from all over the territories of Papua and New Guinea. However, as others have noted (e.g., Bashkow 2006), this sort of change provoked a major shift in Papua New Guinean perceptions of the world.

This historical encounter informs contemporary ecological relationships between persons and places, which are revealed through temporal and spatial layers of myths and stories—histories that are both personal and fantastic. Two of the many layers that can be traced through the sediments of Biangai pasts are cannibals and gold, tropes that dominate both local and global dialogues about Papua New Guinean landscapes. The differing intersections of these two not only shaped early colonial relations and the development of the Bulolo Valley by Euro-Australian settlers but continue to inform local ideas about production and future opportunities. Ultimately, the contradictions in their articulations provide spaces for ongoing expression of Biangai agency and ideas of place.

Various Spanish and Portuguese explorers named one or more of the islands under the assumption that gold would be found among its alluvial deposits. In 1528, for instance, Alvaro de Saavedra dubbed what are now known as the Schouten Islands, along the northwest coast of New Guinea, Isla de Oro (Island of Gold). Later, Inigo Oritiz de Retes claimed the mainland for the Spanish monarchy, calling it Nuevo Guinea.[1] However, many found the mountainous jungles and its inhabitants inhospitable, and these islands became more widely known for their "stone age tribes." Over the years, in spite of its abundant production of mineral resources, highlighted by international court cases (Ok Tedi, for example; see Kirsch 2006) and internal strife (as witnessed during the Bougainville civil war over benefits from the Panguna copper mine) and the presence of some of the world's largest gold and copper mines, Papua New Guinea is still associated with the "primitive," the "unexpected," the stone age, and in its most extreme form, cannibalism. Which raises two questions: Why is mineral wealth so overlooked in the global imaginary of this nation, while images of the primitive persist? How are these ideations emplaced?

Similarly, regarding Venezuela's emergence as a modern postcolonial nation, anthropologist Fernando Coronil notes: "At the close of the twentieth century, Venezuela is commonly identified as an oil nation. Strange as this may seem, a mere material commodity serves to represent its identity as a national community. The remarkable fact that this rather common manner of identifying a neocolonial nation by its major export product seems unremarkably natural only highlights the need to understand why some nations have become so bonded to some commodities that they have come to be identified by them" (1997, 67). Such national images require a certain historical amnesia about how they come into being. The particulars are forgotten, giving more authority to the ideal of an oil nation. Central to this merging of the political body (the Venezuelan state) with the natural body (oil) is the "domestication of value," where a commodity is converted into common currency and permeates society as exchange value. The "primitive" state of Venezuela is thus converted into an oil state, as "oil money" permeates all transactions. Other examples exist, both in the subsoil with South African diamonds and on the surface with crops such as Brazilian or Columbian coffee.

Papua New Guinea, as a national and global image, raises an alternative question. Mineral wealth permeates the national economy, and while not evenly distributed, mineral exports have accounted for 70 percent of the GDP since 2000; gold alone accounted for 40 percent of the exports in 2011 (Banks 2014). At independence, the now defunct Bougainville Copper Mine was imagined as the future of the country. Today, numerous gold and copper projects throughout the nation continue to drive local aspirations, as they connect indigenous places with global flows. So why hasn't Papua New Guinea become bonded to its rocky commodities? Why are places rich in gold and copper and inhabited by incredible biodiversity persistently viewed as places of primitives? Following Coronil, answering this question requires us to consider under what historical conditions the nation's mineral wealth enters into the national and global imaginary.

Cannibalism and gold remain powerful images for Biangai as well. Both might be glossed as forms of consumption, transforming the social and physical landscape into particular kinds of sociality that is extracted from the world. In some ways it is not surprising that cannibalistic images are entangled in discourses about gold mining, as both are associated in Western discourse with the greed and violence of an "other." The political and natural bodies merge together in the national imaginary of Papua New Guinea, linking the production of mineral wealth with the production of supposedly primitive places (see West 2012). In contrast to Coronil's national focus, I consider these as locally embedded discussions about economy, ecology, places, and persons, as extractive metaphors come to dominate relations with place. In the end, the value that is produced, domesticated, and circulated is not mineral wealth but a hybrid of gold and a naturalized other (Escobar 1999). In the place of pure gold, we find cannibal gold and the stone-age tribe.

THEORIZING THE CANNIBAL

It has been noted that cannibalism is one of the most extreme ways to create a moral distance from an "other" (Arens 1979; Goldman 1999). Within anthropology, cannibalism continues to be a much-debated topic (Lindenbaum 2004). However, the focus is not so much on the interpretation and local meaning of such practices as it is on whether the practice ever existed. In spite of numerous anthropological discussions (e.g., Brown and Tuzin 1983; Goldman 1999), no anthropologist can claim to have witnessed it firsthand. Furthermore, it is often asserted that eyewitness accounts by missionaries, exploration vessels, colonial officers, and others are highly questionable.[2] Instead, these accounts can be seen as rooted in colonial ideas of the other, establishing the ultimate in ethnographic difference and granting moral authority to colonial representatives.

In an extension of his ongoing dispute with anthropologist Marshall Sahlins (1985, 2003), Gananath Obeyesekere argues that "the discourse on cannibalism tells us more about the British preoccupation with cannibalism than about Maori cannibalism" (1992, 691), or Fijian cannibalism (1998, 2001), or we might add, Papua New Guinean cannibalism. Certainly, there is merit in this point (e.g., Brown and Tuzin 1983, 3). In Papua New Guinea, for instance, is it mere coincidence that cannibalism coincides with the demarcation of "uncontrolled areas," follows gold prospectors inland, and vanishes once the colonial administration is established? However, in our zeal to deconstruct the historical record, we cannot ignore local conceptualizations. As

Laurence Goldman argued in his introduction to a 1999 volume on the subject, "Imaginative literature and sacred history are mutually implicative, and mutually referential dialogues, not polarized fields of symbolic reasoning" (Goldman 1999, 3). In what ways do different ideas of cannibalism, both indigenous and colonial, "speak" to one another over space and time? The Biangai history of cannibalism and mineral wealth (particularly gold) illustrates how one came to stand for local desires and national imaginings, and the other did not.

For early gold prospectors, fending off cannibals was a central trope of the experience. As they succeeded in conquering the landscape, winning gold from its surfaces, they also succeeded in protecting themselves from the people whose lands they treated as "waste and vacant." When early miners first entered Biangai lands in the 1920s, before there was a town called Wau or creeks and streams named by prospectors, they positioned themselves in contrast to the local population, whose supposed penchant for consuming human flesh was well advertised in the Australian papers. Accusations of cannibalism were part of how capital justified the separation of Biangai from the lands that prospectors mined. For many Biangai today, cannibalism and gold continue to be pervasive themes in their development imaginaries. The local discourse around capitalist development, and especially gold mining, places Biangai stories of ancestral cannibalism as in the way of contemporary development (gold mining, coffee plantations, and the like). Polluted by cannibalism's transgressions, the land and its spiritual occupants refuse to be productive. The relationship between these images of gold and cannibalism forms the global imaginary of Papua New Guinea. But more specifically, it is entangled in Biangai relations to place.

Biangai places continue to be shaped by ethnohistorical accounts of a cannibalistic past, initially formed in the colonial encounter where the land was cannibalized for gold. One line in this cannibalistic imaginary emerges out of the personal experiences of an early prospector, recounted by Doris Booth in her travelogue *Mountains, Gold and Cannibals* (1929). This discussion leads into an analysis of a series of events collectively known as the "Kaisenik Killings," which represent a shift in the relationship between miners and Biangai, as the colonial administration engaged in punitive expeditions to bring the accused cannibals under control. Another line follows contemporary Biangai engagements with cannibalism and gold, highlighting the ongoing significance of this imagery. Mining, it is argued, continues to cannibalize the land as it extracts gold from the very body of their past.

Australian prospector Arthur Darling discovered gold in the small streams around Wau in 1910.[3] Although he died before he could return to stake out his claim, he reportedly told his fellow countryman William "Sharkeye" Park of his discovery. Park and his partner, Jack Nettleton, eventually found Darling's discovery sometime before 1921 (Waterhouse 2010, 19–24). Because of the remote location and the relative size of the find, a major gold rush was held in check (Healy 1967, 17). After William Royal and Dick Glasson's discovery of the much richer Edie Creek and Koranga deposits in January 1926, the Australian press began to advertise the rich but difficult field in central New Guinea. By June 1927, 250 white miners and 3,250 coastal laborers had invaded the Wau-Bulolo Valley (Waterhouse 2010, 56). The Biangai now recall how their ancestors were overwhelmed by this influx of outsiders because it challenged how they understood the world.[4] The appearance of whites, their laborers, and the resources at their disposal revealed a much wider geography and differences in production and resource values that Biangai continue to manipulate.

In the early days of gold exploration (1920s), Biangai were more than reticent to become carriers for the growing mining community, demonstrating a reluctance to work (Booth 1929; see below). Many were in direct conflict with miners and their indentured laborers over the theft of produce and pigs from Biangai gardens (Willis 1977a, 1977b). Gold miners, however, conceptualized their own struggles in two related ways: (1) they struggled to maintain successful mining operations in what they viewed to be the harshest and most wild of conditions, and (2) they struggled to protect themselves from those whom they believed to be cannibals.

Numerous accounts sold the ideas of cannibalism and gold through travelogues and feature-length newspaper articles. Doris Booth's *Mountains, Gold and Cannibals* (1929), Frank Clune's *Somewhere in New Guinea* (1952), Edmond Demaitre's *New Guinea Gold: Cannibals and Gold-Seekers in New Guinea* (1936), Ion Idriess's *Gold-Dust and Ashes* (1933), and Lloyd Rhys's *Highlights and Flights in New Guinea* (1942) provided the Australian public with nationalist images of their own kind making good in the territory's goldfields (Dixon 2001). Idriess and Clune, in particular, had already gained fame detailing life in the outback of Australia. Such tales were peopled with heroic Australians and "primitives" rightly dispossessed of their land, in the perspective of these writers (Strauss 1998). Turning their pens toward the territories of Papua and New Guinea required little modification. For Clune, it was simply another opportunity to celebrate Australian enterprise (Johnston 2017). Others, like Booth, followed many of the same literary tropes found in this Australian bush realism but relied on their own experiences for the narrative. Like many historical texts, these tales lacked firsthand accounts of cannibalism as a practice but viewed the inhabitants of the valley as cannibals all the same. In doing so, they naturalized them as part of the wild environment from which prospectors hoped only to acquire gold. While connecting Biangai persons with places, they did so under the auspices of dehumanizing and denying agency (Stella 2007). For the audience, including would-be miners, they described the possibilities of gold and the means of its acquisition, creating a "spectacle" that engaged individuals and investors alike (Tsing 2000). As Anna Tsing (2000, 118) notes concerning resource speculation, financial capital "must dramatize their dream" to attract investors, and prospectors along the Bulolo were no different. Often sensational, they provided a personal view, a narrative perspective on prospectors' relations with the Biangai.

MOUNTAINS, GOLD, AND CANNIBALS

Doris Booth, the first female miner and nurse to enter the Bulolo Valley, recounts her unique struggles against the environment and resident "cannibals" in a travelogue provocatively titled *Mountains, Gold and Cannibals* (1929). Having remained behind in the coastal town of Salamaua while her husband, Yorkie Booth, went ahead to stake a claim, she led a group of carriers along the steep mountain paths up from the coast into the newly claimed territory. In 1923 she came upon the village of Kaisenik after an arduous trek up from Salamaua. Along the way, she had struggled against inclement weather, rough terrain, and carriers who were reluctant to help her complete the journey. Some were portrayed as lazy or uninterested, but others, she tells us, feared cannibalism. In particular, she describes one indentured laborer whom her party encountered as he was returning to the coast from Wau, quite ill. She kept him around to treat his illness, while he told her "the most gruesome details of how [those ahead] would sneak up on [her] in the night, and of what they would do when they captured [her]" (1929, 73). Cannibalism entered into the discussions of the mining community, but this fear was also shared by those they employed. Assessing the veracity of such claims is not easy, since no carrier authored a best seller for Australian audiences. However, from ethnographic accounts (such as Rumsey 1999), we know that this particular form of stigmatization is as common between neighboring language groups in Papua New Guinea as it was in the colonial setting.

Having accomplished the arduous climb up the mountain and arrived in the Biangai village of Kaisenik, she encountered her first "cannibalistic" image. Up to this point in her account, cannibalism had mostly been a topic of conversation, but here the stories and personal experiences would become entangled. Noting that Kaisenik villagers had recently engaged in warfare, she describes the village as she witnessed it:

> They had been at war a couple of days before our arrival; and there were trophies—gruesome ones—lying around their houses. I saw bones and skulls lying all over the place—without doubt the remains of a cannibal feasts! Some of the houses were decorated with these terrible souvenirs of the chase, and of the feast that followed. Here there was some attempt at displaying them, but in some of the houses the bones were just thrown on the floors. In one house a kind of shelf was erected to support a row of grinning skulls! (Booth 1929, 72)

Later, having been offered a sample of this feast:

> I determined to see it, and Usendran [her assistant] went off with one of the interpreters, and returned with a portion of a human foot! Proudly Usendran displayed the exhibit. I looked long enough to see that the skin was crackled on the foot, and that it looked like a piece of pork.
>
> Then I began to feel very queer, and I waved Usendran away, telling him to bury it. Usendran looked upon it as a huge joke. He had been a cannibal himself, and there was a possibility that the gruesome foot was "buried," but not in the way that I had ordered. (Booth 1929, 73–74)

Booth's training as a nurse gives a certain degree of credibility to her assessment of human remains, and Biangai today corroborate such tales, describing enemies in warfare, such as the Watut, as "our food." One man described to me the practices of his forefathers as they encountered Watut men for the first time, killing them and returning home with the head and long bones for consumption in the village. He had not witnessed or participated in such acts. In all my interviews, no one had. But they all believed strongly in this account of the past.

Cannibalism remained a powerful image for Booth throughout her stay. Having arrived at her husband's camp, she wrote of her struggles with carriers, the relationship she established with some of them, and the events that peppered the early days of the Bulolo gold rush as she tried to maintain a household and tend to the sick.[5] The monotony of working the goldfields was occasionally interrupted by gatherings, typically around Christmas. During one Christmas dinner, she and her husband played host to a group of five fellow prospectors. As the guests gathered, "one item of local news filtered out to the kitchen was that one of our boys, who had been missing, had been killed by our nearest neighbors, the local Kanakas. Probably they had been short of a Christmas dinner!" (Booth 1929, 119). This story was never confirmed, but absent laborers were seen as incapable of easily running away to distant lands, and entering the local food chain was somehow a more plausible explanation. Cannibalism always retains distinctions between known and unknown, self and other. Their "boys" were never associated with such practices (though certainly it was considered a part of their pasts), nor were the laborers of other prospectors. Instead, the cannibal always belonged to the local communities, on whose land Booth and her husband mined. Contemporary residents of the valley often accuse these early prospectors of taking from their land, just as locals were accused of taking from the flesh. As such, places and persons were entangled in a multifaceted exchange. The Adelaide *Register* quoted Booth as remarking that "such a lot of nonsense is written about New Guinea by would-be adventurers that if one told the things that are really true it would only sound like another tall story" (*Register* 1928).

Later, during her first trip to the Edie Creek area, just after the discovery by Royal and Glasson in 1926, Booth took in the scenic view from atop the mountain path: "Here and there the carpet of green or blue was broken by these evidences of Kanaka habitation or industry. Now and again an ant crawled across the picture—a Kanaka or his Mary walking from hut to hut, or some primitive husbandman homeward plodding his weary way—perhaps with his neighbor's head in his bag!" (158). The Biangai, whom Booth credited with some industrious traits, led weary lives, eking out a difficult living but always likely suspects of cannibalism. They were part of the natural landscape that Booth witnessed, working in their gardens and casually eating their neighbors. While scant evidence exists regarding the degree to which cannibalism was practiced, the images were quite powerful and readily consumed by Australians and others imagining the wealth to be won from inland New Guinea. It justified the violence committed against Biangai communities in later years.

Booth, of course, was not alone in her portrayal of locals as savages. While some viewed the Morobe fields as exemplars of modernity, many wrote more adventurous accounts of the goldfields, and the public consumed these tales with great interest. The daily news offered a slightly different line into the cannibalistic discourse of the day. Print media may form a collective imaginary, a shared sense of identity extending beyond the realm of known social relations, formed particularly out of the relationship between print media and capitalism (B. Anderson 1991). In the press and narrative accounts, the Bulolo River and its tributaries were situated in a particularly harsh landscape.

As Booth's struggles up the mountain revealed, the trek inland from the coast required generous stores and the constant recruitment of labor. According to newspaper accounts, both of these essential items were in short supply, and potential miners were cautioned against striking out unprepared. The trek itself took "at least six days" in some accounts, eight for others, always over "razor back hills and dense jungle" (*Sydney Sun* 1926b). Though the mining community was only 6,800 feet above sea level, reporters noted that the route required "three climbs, one of 5,000 ft., and two of about 7,000 ft" (*Register* 1926b). Those returning from the field emphasized to the Australian audience the small size of the area in question and the inaccessibility of the ore, suggesting that it is not a "poor man's field" and that it might be best operated by "a large capital investment" such as a "dredging property" (*Register* 1926b). But these occasional cautionary statements get lost in texts and headlines describing the area as the "New Eldorado," creating a "spectacle" that engaged individuals and investors alike (Tsing 2000). The *Argus* reported: "'I have never heard or read of anything anywhere of such phenomenal richness,' said Hebbard. The owners had not been in any hurry to send their gold out, and at least 50,000 oz was stored on the field, probably awaiting better and safer transport. 'Judging from what I saw,' added Mr Hebbard, 'the field was yielding at least 1,000 oz a day on the labour of about 100 native "boys"'" (*Argus* 1926). In the Adelaide *Sunday Mail* (1926), Hebbard later suggested that while the field might be a limited resource, "there is a big chance of Royal, the man who discovered Edie Creek, becoming a millionaire." Many saw their own futures in this story, and quite a few made the journey. There was another significant part of this landscape that the Australian press could not simply ignore: the human inhabitants of the area. Booth's account of her travels would be published some years later, but the images related in the press seem quite similar. Often it is unclear whether the groups referred to are Biangai or neighboring Buang and Watut (the Kukukuku described by Blackwood [1978]). One article reported: "The natives on the coast were civilized and those in the vicinity of the recent finds peaceable, but farther back they still lived as cannibals and were dangerous" (*Sunday Mail* 1926). In spite of such generalizations, there was little evidence of inland conflict. Disputes had emerged during the initial gold rush of 1924, but they were resolved without violence after a government patrol identified the source of the problem. Relations among the miners, their laborers, and the Biangai seemed to have found a peaceful medium. As a result, the civilizing mission paralleled the development of the goldfields. Aside from concerns about supplies, staking claims, and claim jumping, the availability of indigenous labor and the lifestyles of local populations were significant news items. Many articles reported on the progress made toward pacification and focused more negative attention on the harsh road and life that prospectors must be willing to endure. Reporters often viewed local community members as becoming civilized: "Of big build, over six feet tall, instead of hunting his fellow men for the pot, the native is now becoming civilized and sublimating his energies into working for the white men and earning white men's tobacco, more precious to the nigger [*sic*] than gold, as the fragrant weed may be exchanged for a 'Mary,' as he terms his wives" (*Melbourne Tribune* 1926).

Beyond a reluctance to work (Booth 1929), there was little evidence of hostility. As the *Register* (1926a) noted, citing a recently returned prospector, "The natives in the present known gold areas were quite alright, and gave no trouble if properly handled." Such assessments contrasted with others that deplored the general conditions of the country and imagined this nation as full of savages as well as gold. The image created in the initial phase of the gold rush ambiguously portrayed the local populations sometimes as a threat and sometimes as a docile part of the mining community. Much depended on the purpose of the article, but all maintained that the Biangai and others in the nearby valleys were quite distinctly different from the mining community. The primitive image of the other sold papers, but it also discouraged some prospectors. More importantly, combining a dangerous landscape and a naturalized other helped to separate the gold from the inhabitants.

THE KAISENIK KILLINGS

The images that engaged early miners were heroic: solitary individuals who overcame great obstacles posed by the environment in order to find gold.

Cannibalism fit well into this narrative, as another challenge to conquering the mountains and winning the gold. The miners had to overcome that most extreme of others. The final lesson instilling the dangers of cannibalism in the hearts and minds of the white community involved the events that came to be known as the Kaisenik Killings. Widely reported in the press and a common feature in the travelogues of miners, the Kaisenik Killings remain a significant event for contemporary figures in and around Wau and feature centrally in Biangai discussions of the arrival of whites.

Conflicts began to emerge as small-scale miners rushed into the valley, finding their way along indigenous paths through Biangai villages. Assistant District Officer Appleby reported that during a 1924 patrol to the goldfields, Biangai complained of thefts committed by carriers.[6] According to Appleby, owing to the fact that the few prospectors on the field were seasoned hands in other parts of the territory, they quickly responded with compensation and assurances that quelled Biangai anger. Friendly relationships were ensured by establishing trading ties. He later reported that no further trouble took place at the time.

The gold rush of 1926 was quite different and further shaped perception of the valley as not only rich in gold but also full of cannibals. Following the announcements from the field throughout the press, several hundred prospectors of various experience sought their fortune. They made their way to Rabaul, and then by ship to Salamaua, where they acquired the necessary supplies and the laborers to carry it all up the mountains from the coast. As over three thousand indentured laborers passed through the Biangai villages of Wandumi, Kaisenik, Pinaleng, Salankora, and Lambaura (now Wandumi, Kaisenik, and Biawen), the coastal laborers stole from and trampled on Biangai gardens, often taking pigs and produce as well as cutting down betel nut and pandanus trees (Willis 1977a, 1977b).[7] These problems resulted in violence between December 1926 and January 1927, when a group of three Biangai brothers responded by attacking and killing two carriers from coastal Waria. In the events that followed, more carriers were attacked, three were killed, arrests were attempted, police reinforcements and colonial officers were called in, Biangai villages were burned, their gardens were destroyed, and an unknown number of Biangai were killed. While Booth had tried to purchase liberty for the Biangai, the violence of 1926–27 would forcefully bring them under colonial control.

Reports of these events vary greatly, but like so many other incidents throughout the territory, it underscored the dominant moral message. Booth's account briefly reported that "the local kanakas had killed and eaten five indentured labourers" (1929, 161).[8] Even more sensationally, she writes of the eventual arrests, during which, after losing a number of men to the fighting, "the cannibals were found cutting up their own dead, ready for the table, when the expedition returned to collect the bodies for burial." "The cannibals," she explains, "did not believe in waste" (162). Hers is the only account of endo-cannibalism that I know of for the area. This might reflect rumors about postmortem treatment of the body more than actual practices (Halvaksz 2003).

Idriess (1933), noted for his fanciful rendition of history, speculated about messages sent along the "bush wireless," meaning drummed messages.[9] In one of his many moments of speculative zeal, he quotes the bush wireless messages: "The bush kanakas are short of meat!" (151). And later: "Plenty of meat is coming—take it!" (152). Cannibalism was not unique to this event, he tells his reader: "The whites knew that the carriers who disappeared could not all be deserters, for among them were men from overseas. These had nowhere to desert to. If they took to the bush they knew their certain fate" (152). In reference to the events that provoked such outrage, his account tells of three headless men found by Sergeant Bourke, "their bodies quite warm" (154). Before he could return with a larger party, the bodies had vanished: "The cannibal men came eagerly back and snatching up their victims, carried them to the salt water springs quite near the track and there cooked them" (154).

This history of colonization is by no means unique, as punitive expeditions were a common practice of colonial governments (Keesing 1992). However, what these early encounters say about Biangai personhood and the individual's relationship to the larger group is significant. In his account of these events, Idriess attributes the initial killing of carriers not only to the three brothers but more importantly to the women of the village: "How the young men were taunted by the village women! Sullen within their quarters they listened to the jeering women passing by. A young girl would slip off her lap-lap and throw it over the fence shrilly calling, 'Keep it, brave warriors. I have no use for it now!'" (Idriess 1933, 168).[10] This account is corroborated in the official report of Assistant District Officer Appleby. While men most certainly carried out the violence, in doing so, their agency remained partial or dispersed among the wider set of relations that defines Melanesian personhood. Such persons are compositions of the social relationships that engender their growth, development, and ongoing relations of work and exchange (M. Strathern 1990). Thus, Biangai connections to each other, through the land, remain central. In this case, their wives and female kin, whose labors and resource rights were equally violated by the plundering carriers, entrusted them with the responsibility. As the violence escalated, many often-separate Biangai groups became involved, expressing solidarity with neighboring communities and connecting to one another through kinship and shared connections to place. In contrast, once the violence had subsided, Appleby, the officer charged with bringing a resolution to the conflict, demanded that only the three brothers be brought to justice. For the government, responsibility was not shared across the communities, and therefore it was attributed only to the accused individuals.

With respect to the first attack against carriers, the media were relatively silent, almost casually reporting the deaths. One of the earliest accounts reported, "A radio message received today from Bulolo stated that since the Nak-Anai horror, news has been received at Edie Creek gold fields that the bodies of two carriers had been found shockingly mutilated in a ravine near a small native village. . . . [O]ne body had been almost beheaded" (*Advertiser* 1926). The account follows Mining Warden McLean's description of events. Media coverage of the January deaths initially followed the official press release, which does not mention cannibalism, reporting only that "the Warden said that the village natives are of a very low type."[11] However, more sensational accounts appeared as the violence threatened the future of the area. The *Daily Guardian* (1927) headline read, "Cannibals Attack Porters on Govt. Works," adding, "The bodies were shockingly mutilated, as all that was found was portion of stomachs and shoulders. Evidently the rest of the bodies had been carried away, probably to provide a cannibal orgy and sing sing." Again, while in contradiction to official accounts, the image further connects the goldfields with cannibals. Other elaborations, while less sensational, achieved a similar portrait of moral superiority, calling the population a "vindictive tribe" (*Pictorial Sun* 1927), while most echoed the official media release, noting that the natives were of a "very low type" (e.g., *Pictorial Sun* 1927; *Register* 1927).

In the years that followed, their men's houses were systematically burned (see also Martin 1992), including the central men's house located near Elauru, ending what Biangai now refer to as an independent Biangai government (Haus Parliament bilong mipela). Today, Biangai describe these events as warfare between themselves and the white community, which had not so much alienated land as it had subverted Biangai political authority and subsequently redefined their relations to each other. Within ten years, the troublesome villages that became Kaisenik were no longer thought to be a threat. A practice once described as natural and engrained had simply vanished. In a short article in *Pacific Islands Monthly* titled "Kaisenik: Model Village in Place of a Cannibal Den," the author recounts a visit "with the object of making a mental comparison with the past" (1936, 15). Having reiterated how "natives incurred the wrath of the miners by killing and eating their native carriers," the village was described as completing a moral transformation.

> The new village lies in a hollow, alongside a tributary of the Bulolo River, and is surrounded by native gardens of sugar cane and kau-kau (sweet potato). Pigs run about the village and rear the numerous offspring under the grass roofed houses. The staccato crow of a domestic rooster would indicate that the love of poultry is not confined to the American Negro. The peacefulness of the agrarian scene is completed by the village church, which is the largest building in the place.
>
> In what was once a cannibal village not a warrior remains. The warriors that might have been are away working on the gold claims of their conquerors. The village is in the keeping of the old men, women and children. (15)

The work that was begun by Doris Booth had now been completed, as indentured labor was now the rule and not the exception. For the colonial mining community, they had succeeded in creating new laboring subjects out of cannibals, while Biangai experiences are more complex. Here in its absence cannibalism continues to create images of the other, transformed not by their own agency but by colonial and mission forces. The natural mineral wealth entered circulation not through the domestication of value but through the subjugation of the other. This image persists not merely as the product of ongoing violence in Papua New Guinea. Violence alone does not beget the image of primitivism, particularly when guns, advanced technical and mechanical means for mineral extraction, and an educated leadership and working class are part of the picture. We can see the perpetuation of this imaginary in local development discourses, and in postcolonial labor and compensation practices. While persons and places remained entangled, mining the natural and social body had become a kind of cannibalism.

CONSUMING PLACE OR MINING NATURE

Over almost one hundred years of mining, the Biangai transition from kinship-based modes of production to a neoliberal one has been ongoing (cf. Hyndman 1994). Relative to other inland communities of Papua New Guinea, by the end of the 1930s Wau had become more than an outpost of gold miners and missionaries. The developing township included a movie theater, a swimming pool, telephones, stores, a brewery, tennis courts, a rifle range, motor vehicles, a Masonic temple, a hotel, and regular air service. Euro-Australians entertained themselves with formal balls, musical societies, and visits from Santa Claus. As places were engaged by extractive capital, villagers watched with great wonder and one might even say with great desire. On the edge of this emerging European outpost, the Biangai witnessed a great transformation of their valley, as well as the emergence of gold and timber as valuable resources. Following World War II and the depletion of the valley's easily accessible ore, logging drove the economy into independence. The industries held by New Guinea Goldfields were nationalized into the hands of the newly formed Biangai Development Corporation, including buildings, leases, and equipment. But with independence the market for timber eased and Biangai management of former colonial enterprises proved ineffective. There is no longer a working theater, nor do people play tennis, swim in the local pool, or drink a local beer; however, places remain infused with the practices instilled by this history. Aspirations for managing their own enterprises remain, and Biangai and non-Biangai continue to line the streambeds looking for those last remnants of the gold rush. Relatively little is to be found, and businesses lack the capital and markets necessary for comparable success. The Biangai are left wondering why things changed so dramatically under their leadership.

For the colonial township of Wau, controlling the "cannibals" in the area was central to the emergence of the community and the extraction of gold. Wau's emergence as an administrative center coincided with increased government presence and, most importantly, the advent of commercial flights that circumvented perceived cannibal villages. Today, this same town has fallen into a disordered state, with weekly robberies, large migrant populations, alcoholism, marijuana use, and a state of ongoing violence that lead Biangai to refer to this closest administrative center as a "cowboy town." Here they evoke the images of the American West, as seen on television and bootlegged DVDs. While some debate economic and government responsibilities, others, like the miners before them, blame the lack of development and the regression of the township, as well as their own failures in business, on a past of cannibalism. The same places that were once occupied by settler businesses were also sites of conflict with neighboring language groups, places where enemies were killed and, as Biangai claim today, cannibalism was practiced. Cannibalism figures not only into precolonial and colonial warfare but also into Biangai cosmology. The cannibalistic imaginary overshadows gold and coffee, condemning, in the eyes of some, the township and its environs to perpetual malaise until the past is accounted for, while allowing non-Biangai to prosper.

Those who have established rights in the Hidden Valley project run by Harmony Gold (South Africa) benefit in ways that their ancestors could not. However, in some ways not much has changed in terms of who owns the means of production and who retains ultimate rights to the mineral resources. Under Papua New Guinea mining laws, recognized landowners derive benefits from access routes and compensation for damage to surface resources, and they negotiate a share of mining royalties. Mineral resources remain the property of the government, leased to corporate entities for exploration and extraction. Outside the landowner communities of Winima and Kwembu, Biangai remain on the margins of project development and production. Many work for the mine, but most try to explain their rights, or lack thereof, in terms of Biangai worldviews.

Historical relationships with gold involved both violence and metaphors of cannibalistic consumption that explicitly denied Biangai control. The relationship between cannibalism and gold informs local models of production and relations to place (e.g., Nash 1979; Taussig 1980). The circulation of gold permeates the economy (like oil), but its distribution is not uniform. While some Biangai gain from compensation, others do not; even those who are recognized as stakeholders and work for the mine lack real power over day-to-day operations.

CANNIBALISM AND PRODUCTION

Discussions of cannibalism are not uncommon in the discourse surrounding Biangai views of their past. According to Biangai cosmology, the first man, whose name I cannot reveal, was a cannibal (see also Martin 1992). He was strongly disliked for consuming members of his own family. In retribution, he was killed and buried on a hillside between two contemporary villages, and from his body the first yams were formed.[12] These yams were revealed to his sisters, to whom he had promised to return in a form essential for their survival. As a result, the production and consumption of yams is heavy with meaning, revealing that the ultimate source of productive agency is established through relationships with ancestral gardens. Gardens are spaces that relate and care for Biangai who continue to return to their soils (e.g., Lilomaiava-Doktor 2009; Wendt 1999). Relationships through and with such places are at the center of how Biangai understand failures and successes, whether they involve yams or gold.

In the shadow of ongoing mining for gold, some have questioned the true meaning of the first man's promise to his sisters. While yams remain meaningful, they lack the currency of gold in the new economic relationships. Perhaps something else was intended. Perhaps more was meant in his promise. Perhaps the body of the first man also gave rise to gold. Ultimately, for many Biangai, connections between gold and cannibalism are relatively novel, a product of the colonial. The interpretation that these questions suggest was somewhat common among members of both Winima (inside the lease area) and Elauru (outside the lease area), who do not stand to gain substantially from Hidden Valley's operations or who themselves felt that their rights might be denied. Gold is like the child of an ancestor, I was told by a female landowner from Winima concerned about Watut claims in the area. She composed a song expressing her sadness about the possibility of losing control over the place of the mine, and the likelihood that ancestral spirits would carry the gold away.

Kariri yenge rime rewe Kawia yombu luwa Kilili mowele yabi wak-o.
Yeli Papua ma Wabuni yongo yangu wara wasibu weyame inge nazewi kaiyime ringi wakke.
Yebe wera wizamwe pokko ngairiri bukura kari.
Kari keremiya bera yenge rewe iveni
Yawoa mei Sarowa yarigi yenge ingembu ule rave biyau.
Poromani kiliyi wazamiwe kai iwaremu mize imeng yongebio.
Ngaiyau kopu mopela pongo wizambu.
Nu kiliyengeni kai wizang wi yongo yabi reke wara wiya ingini-o.
[Repeated twice]

[Translation]
The joining of Kairiri and Kaiya, two creeks [associated with two ancestral spirits] near Kilili, marking the boundaries of the Watut and Biangai.
Papuan and Watut women, go and come here as guests, they do not know of what is to come.
The ancestors want work, but watch over with worry.
Like mothers they weep and go to the mountaintop.
Yawoa [one ancestor of this area] tells Sarowa [another ancestor] to look down into the steep valley of the mine

> These two old men look upon the mood of the people.
> They leave with their families going toward the coast.
> Their family [including the gold] collects coconuts along the ocean.
> [Repeated twice]

In this song, the gold is the children of ancestors who inhabit the range now occupied by the mine. These named ancestors could decide to leave with their children if the mining agreement does not adhere to proper genealogies. But this is not simply a metaphor. The ancestors are real agentive forms, placed along the landscape and folded into the mine site. Only those who are children of the land can be allowed to consume such gold.

Part of the way I gathered information about relations with the land was through participatory photography, where community members are recruited to take pictures that are later discussed (e.g., Halvaksz 2008a, 2010). One photo stirred a good deal of controversy. Taken by a Watut man (who was employed to look after the coffee garden of a mine worker), the photo showed his family enjoying a lunch break while washing clothes by the river. In my interview with him, this was all that was discussed. However, when he shared the photo with his Biangai neighbors, the face of a man was revealed in an outcropping of rocks. Many asked to see my copy, and much was made of what it might mean. I tried to play devil's advocate, suggesting that such an image could be explained by the lighting or time of day, but other opinions circulated with greater local credibility. Most common was the belief that the head was an ancestor protecting the land from trespassers, including the Watut photographer. He appeared to let the Watut family know that he was there, "watching," I was told. However, those whose rights derived from a place, an area associated with the Winima villagers who were alienated from primary rights to Hidden Valley, had a different explanation. The area had been subject to gold exploration, and some gold had been found but never developed. A number of households that I frequented concluded that this was no devil but "the head of gold" marking the ground of an ancestor that looks after it.

In Elauru, outside the area of the mining lease at Hidden Valley, they struggled to understand why gold was not found on their lands, and some sought to extend their rights to the gold at Hidden Valley through assertions rooted in this same Biangai cosmology. When one informant described the cosmological link between the first man and his promise to his sisters, it was not yams but gold. Elaborating on this point, he noted scientific data that he had gleaned from conversations with geologists and a brief stint as a laborer on the mine site, rhetorically asking, "Geologists say that gold comes back this way, now where does it go?" The gold, he told me, is not the child of recent ancestors, as claimed in Winima and Kwembu, but the product of the first man, who was buried near Elauru. He argued that this is what the geological data indicated: the head of the gold seam rests beneath their own village. The mine was only operating at his feet.

In this reimagining of the relationship between a cannibal king and Biangai land, the first man promised not yams but a garden—a garden that would attract many and care for all Biangai. My friend noted that gold certainly fulfilled this prophecy better than yams. For an Elauru man, alienated from the Hidden Valley project, reflecting on a history of resource extraction, the promise was for something greater: "He foretold of the gold, and of these pine trees, Klinkii pines and hoop pines; [the yam] was a metaphor." Development opportunities were plenty, and all were rightfully theirs to claim, but none had come to fruition. Gold offered the most potential, as Biangai had witnessed the transformation of their valley under its guidance. However, it had yet to permeate their lives in a substantial and ongoing way, raising concerns about what negligence had been committed, or as one group of village store owners complained, "What kind of sickness do the Biangai have?" The answer most often offered was a moral one: the consumption of flesh by their ancestors held them in check, prevented development, and would do so until amends had been made (cf. J. Robbins 2004; M. Smith 1994).

APOLOGIZING FOR CANNIBALISM

Early one morning in 2001, as Paipe, Chris, and I walked from Elauru to Wau for supplies, we fell in line behind a group of middle school students making their daily trek to the community school in the village of Kaisenik. As we turned the final corner, descending toward the first houses of the village, Paipe took note of a printed secondhand T-shirt worn by a young man in front of us. He tapped my shoulder and pointed, asking if I thought it was a reference to the Biangai. The shirt was typical of the secondhand clothes that enter the global flow of discarded items from Australia, Europe, and America (Hansen 2004). Slightly worn, the fading design on the back depicted a book cover from a classic B-novel. "Cannibals" was the only legible word, with an image of a dark figure looming over a young white teenager, female if my memory serves me. Paipe pressed me for information. At the time, thinking nothing of the image, I explained how teenagers in America favor such clothing, trying to get across the idea of being retro to my friend and research assistant. The image, I explained, was taken from the cover of a piece of fiction, "stori nating." I assured him that it was not necessarily about the Biangai or Papua New Guinea, and it certainly was not a true story. Our conversation faded as we passed through the village, stopping to chat here and there on our way toward Wau. As we walked on, Paipe, still thinking about this image, commented that local Pentecostal church leaders were going to apologize for such practices, and that by doing so they might make development possible.

Plans had been underway for some time to *tok sori* (apologize) to those whose ancestors had died on this land. Pastors wanted to invite Watut communities to feast and make this apology in hopes of restoring the productivity of the land. While this feast never eventuated, the idea remained a powerful way to explain failure and the regression of the township of Wau. Biangai efforts at most things, whether the management of stores and businesses, coffee and other cash crops, or attracting international mining companies to their land, seemed to always fall short, if not fail. At the time of this conversation with Paipe, Hidden Valley was in the seemingly endless phase of exploration, adding to local anxiety. The ground, many felt, was against them as an active agent of past wrongs. Christian emphasis on sin, merged with Biangai connections to place, was hindering both their ability to act and the power of the place to be bountiful. In this case the wrong kind of consumption had occurred. Restoring and strengthening their connections with places would result in a rejuvenated economy, reducing conflicts and the malaise that had settled into the valley.

On a sunny afternoon soon after my discussion of apologizing for cannibalism with Paipe and Chris, a senior Winima man and I were discussing this dilemma of development when he started telling me his version of Biangai history. The ground, he explained, was the source of the problem, because it was on this ground that men died in precontact warfare. It was on this ground that Biangai fought and consumed others. Coffee would never grow there, nor would development come to Biangai who didn't acknowledge this error of their ancestors. He laughed off the effort by yet another company to try to develop coffee in Wau. Like the mining efforts and many Biangai attempts to do the same, it would surely fail.[13] What is most striking about this discussion is the apparent shift from an emphasis on the role of one's ancestors as sources of productive power to an emphasis on severing parts of those connections. Michael Smith had observed similar debates about production in Kragur, noting that "they see moral and material well-being as having a circular and interdependent relationship" (1994, 7). Like Kragur, Biangai increasingly believe this well-being is defined through other sources of morality.

Embedded in Winima and Elauru's imaginings of gold's purpose is a dialogue about their past and future. On the one hand, they believe that the gold at Hidden Valley rightfully belongs to the Biangai (especially the two villages within the lease area, Winima and Kwembu) and not to the Watut. It is produced from the body of the first man or connected to more recent ancestral forces, and the Biangai should prosper from its extraction. On the other hand, gold extracted from the valley during the colonial period was taken from its rightful owners and consumed elsewhere. This is apparent in claims to retribution from the Australian government by groups of local officials (see *Post-Courier* 1980). For the Biangai in particular, it was not merely a theft. When the gold is conceptualized as a product of ancestral agency, colonial gold mining was more than the extraction of resources that rightfully should have benefited the Biangai; it was a cannibalism of Biangai places. The early prospectors consumed the physical wealth promised by the first man, created through ancestral agency for Biangai benefit.

However, failure in this analysis is not solely the product of this exo-cannibalism of their resources, where endo-cannibalism was intended. Attributions of Biangai failures and Wau's decline derive from a reversal in meaning, the product of missionization and pacification (Clark 1993). The positive productive agency that might once have been attributed to warfare and the practice of cannibalism is reversed, giving a negative moral weight to these practices and having a negative effect on the productivity of the land. Ultimately, it requires a reconsideration of relations with ancestors and places. Such connections remain important, grounding Biangai discursively and pragmatically in a future tied to place.

WHY PAPUA NEW GUINEA IS NOT VENEZUELA

In 2009, violence erupted between Watut and Biangai over claims to a mining lease located in McAdam National Park (the lease was technically illegal). A Watut man was killed for what Biangai claim to be trespassing. Following years of anger over the control of Wau Township and the former properties of New Guinea Goldfields, hundreds of Watut and other settlers rioted through the area in response. At first, they destroyed the Biangai settlements on the edge of town, killing two (a Biangai elder and a disabled youth), before moving through the city, raiding and burning as they went. On the second day, they raided and burned Kaisenik village entirely to the ground before turning toward the nearby village of Wandumi, only to be halted by police reinforcements. For the national and international press, this is where the story ended. But for Biangai, a miracle occurred at Wandumi.

According to several who witnessed these events, they used prayer and only nine bullets to hold off an angry mob. As all of the Christian leaders from every denomination sat and prayed in an old church building, a small group successfully defended the bridge into the community against powerful witches and customary magic. For Biangai, this victory was tied to their rights to place, their belief in God, and their rejection of the past. In contrast to earlier conflicts, Biangai saw themselves as in conflict with a more primitive other. The source of the conflict was not simply Biangai control of mining leases around Wau but an expression of Biangai connections to place. Even though tensions remained high, Biangai refer to this as "Biangai Easter," marking a miracle or rebirth of the entire Biangai community. For several years afterward, a coalition of women from all denominations traveled through the seven villages holding revival-like services, staging performances that re-created the events of the conflict, and talking about a shared future for Biangai that would emerge out of their belonging to this place. As one organizer explained to me in 2011, "When we came together as one tribe, one language, and one place to pray to God, this had positive results." At that moment, future social relations among all Biangai were viewed with hope.

Ideas about cannibals and violence closely associated with Papua New Guinea continue to circulate both internationally and locally. But local ideas about such pasts likewise reveal an economy driven as much by questions of morality and the moral value of place as it is by desire for wealth. Biangai lands are moral spaces where cannibalism and violence are not feared practices but past wrongs. While gold's impact on the broader economy of Papua New Guinea is widespread, its circulation remains uneven, highly localized, and extremely temporary as gold rushes and gold mines ebb and flow.

In the villages around Wau, responses to the sporadic presence of gold have always raised questions about resource rights and the meaning of gold. Early miners imagined the "cannibals" around them as potentially consuming their flesh and that of their laborers, should they fail to be watchful, bold, and aggressive even as they delved into the earth's surface to extract gold. The natural body and political body were clearly distinguishable. While the Euro-Australians fit solidly within the political, the natural encompassed mountains, gold, and cannibals. For the Biangai the distinction between the natural and political is less certain, as they imagine a landscape that is ghostly and ripe with spiritual agency, which defines their relationship with places and through places to each other. Thus, the domestication of gold and its circulation among Biangai continue to link the natural and the political body in significant ways, informing how they see and experience place.

Unlike Venezuelan oil, gold's circulation in Papua New Guinea not only permeates the economy of the modern nation but does so still infused with history and local meanings that continue to link it to ideas of the other, including both violent enemies and the cannibal. Gold literally embeds into the land a moral message, and thus the relations among persons. Gold might be domesticated in places like Wau, where it circulates among international mining companies, designated beneficiaries, government officials, small-scale miners, criminals, and communities, but it does not do so unencumbered from rich historical and local meanings that continue to connect it to cannibals, whether real or imagined. Much of this local meaning is formulated in a specific relationship with the land.

CHAPTER TWO

Grounding Kinship

As he often did, my host father came for the weekend and headed back to town early Monday morning. Working for the mine as a driver, he had only a short time to spend in the village. His coming and going happened often, such that the emotions of parting were mostly muted. Such partings have become part of life in this remote community, as the mine employs an ever-increasing number of men and women. Because such events are commonplace, when he left in the morning we didn't see him off. Later in the day, when I, along with other members of the house, complained of being tired, Mama Nawasio explained that it was because of our father's *imeng* that we were tired. *Imeng* is hard to translate. It can reference one's disposition or part of one's personality. Some said it was one's "dirt" (tp: *pipia*), but they felt that that wasn't quite right. Either way, it is a part of oneself that remains behind, tied to the place through social relations. Mama Nawasio explained that when a close family member leaves the house, that person's *imeng* makes those of the same house tired or even sick. The first day is the worst, *imeng ngaira* (powerful *imeng).* At death, *imeng* hangs in the air of the house, as the family experiences a great sadness. It is quite literally an agentive part of oneself left behind that can affect the emotional and physical state of others.

Local idioms of relatedness are embodied in the relations of placepersons and the physical landscapes Biangai work on a daily basis. In simplest terms, relations among persons *and* places are defined through kinship. More than mere property, places form parts of persons, and persons are composed of their relations with places. It is within these relationships that Biangai engage in negotiations with mining and conservation, and it is through emplaced kin that they hope to effect change. But what they mean by kinship is quite distinct.

Biangai kinship is linked to the access to and use of resources and the practices of creating and re-creating social landscapes. More than a valley created through the geological processes that mining companies examine and scientists appreciate (e.g., Gressitt and Nadkarni 1978; Dow, Smit, and Page 1974), the Bulolo Valley is surrounded by mountainsides lined with ancestral paths (*kasi mek*) that are traveled for hunting and gardening—routes that lead to other villages, other places, some real and some imagined. The first Biangai, if not the first humans, left their footprints on paths that continue to be followed today.[1] These ancestors could take the form of pigs, birds of prey, and wild dogs, animals associated with named kin groups, and they gave life to the first communities.[2] The mountaintops, the flowing waters, and the ground itself continue to serve as residences for these ancestral ghosts.[3] More than just metaphor, they are imbued with *imeng*. Their heroic confrontations started the landslides that still cut into the grounds of the Samuna Community School, moved the giant salt stone from the head of the Bulolo River to the village of Wandumi, and most importantly, created yams and gold from their own decaying flesh. Their agencies are not lost but remain as powerful forces in hunting and gardening, life and death.

That humans and places mutually create one another is not a novel idea (A. Anderson 2011; Leach 2003). It is central to indigenous scholarship in the Pacific (e.g., Ka'ili 2017) and grounds the approach of a decolonized methodology (e.g., L. Smith 1999; Nabobo-Baba 2006). As places are constituted by the actions of humans, and humans by their actions with places, Biangai worlds are layered with human and nonhuman agencies—agencies that are related to one another through kinship. These relations are most strongly realized in yam gardens. As the ideal crop, yams continue to be treated as children, planted by men who give them their strength but nursed by the attentions of women.[4] However, the right to plant in a given location must be recognized by ancestral agencies in order to ensure yam growth. Yam gardens are joint efforts, planted on the land of a family group to which men and women share rights. They are the offspring of Biangai and their ancestors, products of human labor and ghostly agency, gifts given to them like the animal that gives itself up to the hunter.[5] Thus, kinship links the living and the dead to the environment as individuals trace ancestry across the landscape.

Such relationships are recorded in funerary songs composed to honor recently deceased loved ones. These narratives engage the "soul-stuff" (Machio 1994) or *imeng* of the deceased as it is dispersed across the landscape.

Women compose these songs, called *yongo ingi*, in the days after a death, and they are rehearsed at subsequent funerals as families sit with the body (Halvaksz 2003). A senior woman from Winima composed the song below out of respect for her sister:

> *Kamua yango yoko moha kazei tiyau hurani rua mere*
>
> *Ni yenga imi ni silang waka naize keri nai ipela yana sabiya kuamo oeai i terika*
> *Ruwa/ mobu nala rege sahuwaro yakavuo*
> *ya nalumabu roui hurani yenalea/ mara wairi kaima raya*
>
> *Nimake remu rewe kovori salulabu yorusiki panu mere*
>
> *Ire yo sinibu waima pela*
> [Repeat twice]
>
> [Translation]
> Kamua Yangon and Yoko Moha, their mother died
>
> She left along the creek they call Silang Kosa, she goes to Sabiya to the joining of the two rivers of Kwembu
>
> A landslide where you walk, to the flower where the rivers join
> A boundary mark near Wau Creek, you hold it in your hand and go
>
> Your forest, the cordyline that is found only there, you are the mother of the forest
> You see these things and go
> [Repeat twice]

Like other songs, the lyrics speak poetically about the connections that the deceased have with the land, recording her relations to places and to others who could make similar claims to the named forests and rivers (e.g., Feld 1990; J. Weiner 1991). In singing this song at funerals, women recount relations among the dead and among themselves, but always through places. Increasingly, these songs also evoke connections to mineral rights and compensation, indicating specific places to which the deceased and the composer have rights.

When Biangai debate their connections to places, they speak of their *ngaibilak* (*ngai*: look after, fight for, lead; *bilak*: earth or ground), a term that includes the ground, trees, small streams, and animals that inhabit a given area. Actively looking after the land highlights a subjective engagement with places. While an exact translation might suggest simply "looked-after land," *ngaibilak* is suggestive of the more intimate relations of placepersons (e.g., Māhina 2010; Nabobo-Baba 2006). One enters into "caring-for relationships" (Lilomaiava-Doktor 2009), just as the land, imbued with the spirit of past gardens (and past mines), cares for Biangai gardeners (and miners); it is important that such indigenous concepts be integrated into theoretical frameworks. *Ngaibilak*, then, is not just a local equivalent of a Western concept, but a theoretical position from which placepersons derive meaning. Places and persons look after each other in a Baingai theoretical framework.

In tending to named locations, Biangai distinguish further categories of relatedness, such as *ngaibilak panumere* (looked-after motherland), *ngaibilak mangobek* (looked-after fatherland), acknowledging rights and relationships gained through both one's mother and one's father. Places themselves are not permanently gendered as much as they trace the lineages of persons, embodying the very essence of larger place-based kinship groups called *soloriks*. Rivers and streams form boundaries between parcels of land, marking paths traveled by malevolent male and female spirits (*kauri-bek* for male spirits and *kauri-mere* for female spirits). Ridges and valleys also serve to mark the edges of place-based social relations. But in both cases, these are novelesque spaces, negotiated and contested (Halvaksz 2013a), just as the novel is an open-ended text, subject to multiple interpretations and representing many different voices, in contrast to the epic's "impenetrable boundary" (Bakhtin 1981, 6), which is definitive, set in stone as law. Biangai boundaries are novelesque—dialogically created spaces that are entangled into kinship relationships through song and story. Place boundaries are subject to negotiation, as are relationships among persons.

PLACE-BASED GENEALOGY

The study of kinship has shifted from a focus on politico-jural organizations and lineage systems to localized systems of meaning, as biology (in general) and sexual reproduction (in particular) are not universally appealed to as the basis for kinship systems throughout the world (Schneider 1965, 1984). In Melanesian ethnography, this is apparent in the shift from debates about the use of African models in Highlands New Guinea (e.g., Barnes 1962) to a focus on processes that transmit bodily substances and establish relatedness.[6]

However, more recent studies of kinship have called for a reimagining of the nature-culture divide, as the singular emphasis on the social aspects of relatedness is challenged by contemporary networks that link the material and social worlds.[7] These studies draw on the idea that networks include the material and immaterial, persons and objects, in the construction of hybrid systems of meaning and relatedness (Latour 1993). New reproductive technologies, open adoption, as well as a myriad of local conceptualizations produce persons whose kinship mixes biology, technology, and sociality (e.g., Bamford 2007). Jeanette Edwards and Marilyn Strathern (2000), for example, sidestep debates around biological and social kinship to look at their intersections in ideas of belonging to place, showing how belonging to the town of Alltown in Lancashire is rooted in a sense of belonging to the families of Alltown and its associated patronyms. Relatedness is constructed here both because of belonging to biological families and in spite of biological connections between families.

For Melanesian ethnography, residence and place are also significant sites for rethinking the social, biological, and material basis for relatedness.[8] In response to mining and missionization, a Min ethnic identity has emerged to acknowledge broad-based filiation among otherwise diverse communities and kin groups (Jorgensen 1996). These are neither organized around simple biological kinship nor built upon mythical affiliations, but instead reflect a combination of the two. Likewise, in the process of entification, "contingent categories" are made into entities in the context of resource disputes (Ernst 1999). Such processes are much more extensive than simply a reorganizing principle, and they lead to a process that has been called "culturalization" (Gilberthorpe 2013, 263). Regardless of what we call it, in each case identity and group affiliation are organized in response to development opportunities, reconfiguring kinship in relation to place-making practices. This is what Biangai have always done as they respond to ecological and social conditions of *ngaibilak*. The difference now is that place-making practices are framed in the context of conservation and resource extraction as much as in terms of subsistence gardens and hunting paths.

Echoing Margaret Mead's Arapesh ethnography (1935), James Leach (2003) has shown that such practices are not limited to colonial and postcolonial projects, but appear central in local models of relatedness (see also Hirsch 2004; Jacka 2015). The role of place-making in the making of persons among Nekgini speakers (Madang Province, Papua New Guinea) is not limited to bodily substances or foods as substances; Nekgini-speaking

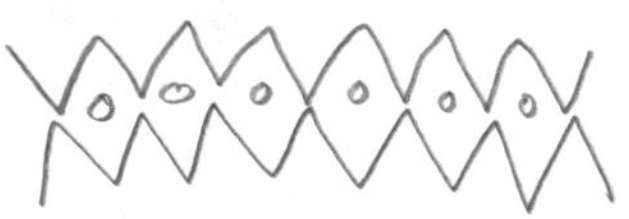

persons are grown because "they draw substance from the land" (Leach 2003, 29; see also Mead 1935). Outsiders can then be made into kin when fed from Nekgini places. This is similar to Hagen clan identity, which allows for the addition of nonagnates and Porgarean practices that facilitate adoptions and sponsorship.[9] Place, in association with agnates and nonagnates, creates relatedness by growing the body of the clan, which is accomplished through reproduction and the planting of outsiders (M. Strathern 1990, 255; O'Hanlan and Frankland 2003). Ultimately, persons are made through the transfer of substance from the bodies of men and women and from their gardens—growing the fetus through "accumulated" blood and sperm, or growing nonagnates through the analogous consumption of garden produce.

Among the Nekgini, this principle is extended. Place replaces bodily substance as the generative principle for growing persons. A parent does not pass bodily substance directly to the child, but instead through work "establishes the conditions for the [child's] growth on the land" (Leach 2003, 30).[10] As with the Biangai, the land is alive and "enters directly into the constitution (generation) of persons" (30). In this configuration, being a Nekgini person results neither from fulfilling the role of a person in a given kinship system nor merely from procreation. Instead, "places enter directly into the generation of persons, while persons—through their work—engender places" (Leach 2003, 30–31). But more significantly, *ngaibilak* suggests an intimacy between persons and land, where the caring-for relationship (*ngai-*) defines both the actions of persons and the role of the ground. This mutuality is important for Biangai world-making practices and remains a source of hope.

Biangai conceptualizations of kinship combine the generative power of place with the work of maternal and paternal substance to grow the household. Persons are not complete at birth but rather are grown through love and support, through the allocation of land, shared labor, and gardening, and through marriages that unite parcels of land. People are like yams, requiring the care of a married couple to plant, tend, and distribute.[11] In contrast to Leach's observations about Nekgini, maternal and paternal contributions of substance remain important. As seen throughout much of Papua New Guinea,[12] men contribute the bones and strength of both children and gardens, while women nurture their growth, adding flesh to the child as it develops. If Biangai persons are made through their relations to place, it is because places too share maternal and paternal connections across an extensive network of kin. There is a cosmological basis for the creation of places, the generation of related persons, and differences in how Elauru and Winima economic engagements shape the work of making placepersons.

"YAMS ARE LIKE CHILDREN"

Throughout much of Papua New Guinea, yams are often symbolically linked to human beings. Inhabiting an underground realm through which they move and grow, yams are classed by Dobuans "with [their] own people as personal beings" while excluding whites from this same category (Fortune 1963, 109). Yams are similarly classified by the Arapesh (Mead 1935, 1938) and also play a significant role in masculine group identities. Beyond their phallic associations, Arapesh yams mirror human genealogy. At marriage, both men and women inherit a set of seed yams from which they will plant distinct crops. This distinction is maintained until the first child is born. For a married couple "eating yams is a mystical as well as physical act, and in sharing their ancestral yams a married couple is engaging in a form of spiritual communion" (Tuzin 1972, 235). The importance of maintaining the distinctiveness of each ancestral source is significantly reduced upon the birth of their first child. At this time "the spirits of their yams have fused" (Tuzin 1972, 236). In this way, yams are "parts that circulate as parts of persons" (M. Strathern 1990, 192), thus emphasizing the hybrid network among persons and gardened substances, and the agentive qualities that yams possess.

Today, Biangai seem relatively less concerned about maintaining the distinctiveness of the seed yams but still emphasize the importance of passing such resources among extended family members. Yams are distributed broadly through an extended network of kin. Men and women often ask their families for seed yams to add to their own. These they plant together in family gardens that likewise evoke ancestral relationships through place. Seed yams mark a genealogical connection to both prior seasons and prior gardens. They fold together present gardens with past efforts and present gardeners with past gardeners. Thus, yams and persons are placed across a landscape that is pregnant with historical significance.

Biangai likewise extend this conceptualization to the successful production of other garden produce, both wild and domesticated animals, trees, and even gold. Such productive efforts can involve magic. Though few now possess such abilities, spells were once used to encourage growth or hinder the success of others. As a result, the successful production of yams is a sign of one's good relations with the living and the dead. Thus Biangai yams and the yam gardens are in many ways analogous to the creation of persons and groups, and embody the principles for organizing sociality through place-based relationality.

Biangai kinship practices might be thought of as cognatic (Martin 1992; Mitio 1981; Ouellette 1987) or bilateral with a patrilineal bias (Burton 1996a).[13] While the distinctions in these positions are subtle, more extreme contradictions can be found in early patrol reports. Patrol Officer Maclean determined that inheritance was matrilineal,[14] though Patrol Officer P. G. Whitehead later observed: "Ownership of land is more complicated in the Biangai than most other areas. A male child can claim ownership of both the mother's and the father's land. The mother apparently has equal rights with those of her brothers though this warrants further investigation."[15]

It seems unlikely that such differing principles could uniformly be applied to Biangai practices. Instead, these classifications underscore the problems inherent in formalist models of kinship (e.g., Anderson 2011, 180). Sometimes, and in certain circumstances, each is certainly true. An argument could also be made for matrilineal tendencies in certain families and in certain situations. For example, the family of a woman who marries a non-Biangai man will rely on the woman's land rights for several generations. To say that they are mostly any of the above not only oversimplifies their actual practices; it ignores Biangai genealogical principles that are not rooted in Western constructs.

Kinship terminology follows a generational model (which Morgan [1871] classically called Hawaiian), where gender-specific terms are used for both maternal and paternal relatives. The generic terms for mother and father (*panumere* and *magobek*, respectively) refer to one's biological mother and father as well as their siblings. The same principle applies to each generation. It also helps to organize kinship rights between *ngaibilak panumere* (motherland) and *ngaibilak mangobek* (fatherland), establishing seniority and leadership in relation to place.

For Biangai, relatedness is expressed in metaphors of growth and land—including both garden and human reproduction. Before marriage, male and female children are assigned primary membership in one of the named landholding groups, or *solonariks*. These might be viewed as cognatic stocks (see McArthur 2000), limited in number but not fixed. Over time, new ones can be formed and others can lose their efficacy as persons realign their filiation within the community. The potential for membership in a given line is based upon the Biangai concept of *solorik*. *Solorik* means "stem" or "base" and evokes the image of the base of a clump of bamboo (*negelebek)* with many branches extending in multiple directions. The *solorik* is not unlike the linguistically similar Kunimaipa system of *kapot* (base or source), which identifies the multiple paths that a sibling group can follow to establish resource rights (McArthur 2000). *Solonarik* refers to one line from that base (the *-na-* indicates one). When describing it, Biangai would hold out a closed fist and raise each finger, repeating "solo*na*rik" as each finger was extended. *Solorik* is also used to describe the yam's multiple vines, derived from a single mound, mounds that are in turn found in gardens shared among a group of closely related kin according to the same principle. Each mound produces a mass of leaves that twine up their individual sticks (*sisi*) before intermingling with others, like a roof over the garden. Biangai liken the resulting structure to that other quintessential form of masculine labor, the men's house. *Solorik* affiliations organized the preindependence men's houses, with the house mirroring the yam garden in some basic elements. Inside the men's house, one corner was occupied by one group, a wall by another, and so on. The posts associated with the spaces occupied were in turn supplied by named kinship groupings during construction. The men's house was also framed by posts connected to kinship of places and persons, just like the multiple vines of the yam twining up an individual stick to merge high above (see also A. Anderson 2011).

More than the name of a kind of group, though it is this as well, the *solonarik* is a principle for affiliation. Individuals find themselves representing a number of *soloriks* derived from each parent and grandparent. One can trace connections through the mother's mother, the mother's father, or the father's mother or father. Tracing out social relations is likened to being at the base of a "clump of bamboo," ascending along the many "hands" or stalks. During funerary events, a taboo prevents cutting certain varieties of bamboo, because like the deceased, the bamboo stalks originate from a common source. Mitio (1981) lists several terms for such family groupings, including *iwi wamenak* (one blood) and *yeri namek* (one descent), which are somewhat interchangable with *solonarik*, indicating the role of shared substance in determining affiliation. However, *solonarik* groups do not have to follow a specific matriline or patriline, can skip a generation, can change over the course of one's life, and can switch between patrilineal and matrilineal descent lines between generations. Hence *solorik* groupings are specific paths within an extended multigenerational family. Only the sibling group (including classificatory siblings) represents a common base in this ego-centered network (Ouellette 1987, 104–5). As Ouellette observes, "Each person has a *solonarik* different from all other persons except in the case of unmarried siblings. Thus, each *solonarik* blends several other *solonarik*, and its members are spread across Biangai territory. Members of the same *solonarik* have obligations to one another, more or less restrained by their degree of relationship. They must provide assistance in conflicts and are not allowed to fight one another when they live in two warring groups" (105; my translation).

The affiliations that *solonarik* paths produce are loosely formed into named kinship groups of varying sizes and associated with specific places and paths through the surrounding valleys and mountains. These family groups hold title to garden lands, hunting paths, and specific house sites within the village. Individuals with rights to these areas establish their connections not only through genealogical claims but, more importantly, through a detailed knowledge of the landscape's history. Such histories are spatially represented; knowing who worked a parcel of land and investing it with part of themselves allow the Biangai to create a hybrid of space and human biological and social relations. Thus, one works a particular area not only because of shared substance with kin, but also because of shared substance with the land and the ancestors who worked the land before.

Each contemporary village is primarily composed of at least two or three large *solorik* groups, but it is not uncommon to find representatives of all the named groups in a single village. Three primary groups exist in both Elauru (Kaiwi, Yaru, and Yauromura) and Winima (Igulu, Kaigowi, and Paro). However, numerous subgroups form as village conflicts and marriages shape community relations. The subgroups are often associated with named animals, such as pigs, sea eagles, or wild dogs that are said to occupy an area. These are characterized as "paths" (*kasi mek*) between and within the larger groups that are often defined through the very real hunting paths shared by members. Similar Foi paths in the Southern Highlands of Papua New Guinea have been described as creatively shaping the earth: "They transform the ground, partition the earth, and create human space" (J. Weiner 1991, 38). Paths mark the creative and flexible relationship between people and the land.

Residence becomes central in delineating the relationship between persons and places, as a classificatory sibling will "forget" or "give up" rights to a garden when marrying or being adopted into a distant village. There, he or she will claim rights in another *solonarik* path of parents or grandparents, finding parcels that make sense given the history of land and labor. In contrast to usufruct rights, the new residents become full rights holders in the parcels. Use and residence place persons into social relations by delineating primary rights and organizing decision-making regarding garden lands and, more recently, compensation. In the end, it is the genealogical story that one can tell about connections to land that places persons into kinship groups.

ADOPTION

The variability of group designation is highlighted through adoption (Brady 1976; Goodenough 1970). Often, rights are designated through adoption or naming practices, where children are named after and encouraged to visit, if not live with, male or female kin (see also Burton 1996a). The relations among natal and adoptive families are strengthened through this movement of children, because they maintain hereditary social relationships (Barlow 1985, 2001) and are literally parts of other persons (Lipset 1997, 60). A Biangai child can be adopted between villages, returning the child to lands given up by someone with significant genealogical connections to the adoptive place or in order to activate the rights of a childless couple. In this sense, adoption is as much a practice of place-making as it is an expression of relatedness between families.

With rights dispersed across villages or a *solonarik* group, children are sent to "look after" the land of their mother or father. Historically, it was common for a daughter to be sent to marry in another village, both to strengthen political alliances and to take up rights that one of her parents might claim there. She in turn would send a child (preferably but not necessarily a female) back to her natal village to make similar claims (Biangai refer to this as *yereng solorik mizamani*, or "giving back to root place"). Adoption and marriage are common ways that this is accomplished.

The significance of adoption in organizing families was highlighted in early patrol reports. For example, Patrol Officer P. G. Whitehead complained that the frequency of children's movement between villages made his census work most difficult, as they were often listed twice, in total agreement, by households in different villages in spite of Whitehead's insistence on defining a single residence. While this inflated census numbers and frustrated patrol efforts, the practice also maintained *solorik* affiliation between villages and in important ways resisted the administration's attempt to formalize what was often incorrectly assumed to be a clan system. However, the *solonarik* allows for flexibility within group organization as persons respond to different conflicts, resource needs, and more importantly, to marriage opportunities. In short, it allows for an ongoing calculation and creative imaginings about relations among places and persons.

SECONDARY AFFILIATIONS

While an individual is designated for primary membership in one of the named groups following the principle of *solorik*, he or she retains secondary rights in all of the other groups associated with the parents' and grandparents' generations. The primary affiliation is typically given by the parents as each child nears adulthood, if not before. Parents attempt to designate at least one child to each of the groups that they can claim, ensuring that these rights are not lost. As one man explained to me, parents must "show that the *ngaibilak* [landowner] is present. . . . Otherwise another might take it." Use rights remain in all of the *solorik* affiliations associated with the family of procreation, but other siblings will "go first" in watching over this land. This flexibility is especially salient during droughts, when landslides impact wide garden areas, or when invasions of wild pigs threaten crop productivity. In all cases, consensus must be sought among the extended network of senior relatives before primary rights are granted.

Secondary rights do allow one to speak in meetings regarding an area, to protest exclusion, to hunt, and to defend boundaries. When a primary holder of a given area begins a new garden, he might invite those with secondary claims to plant with him. Not doing so can result in tensions within the family and disputes about primary control. This is especially true of the large yam gardens that families produce together. For example, when one Elauru man began cutting secondary forest to prepare a yam garden during the 2001 season, he failed to immediately notify all those with some claim to the garden area. Even though he had primary rights and was acknowledged as "going first" in this particular land, a female relative was a bit miffed at her exclusion. Was not her father's father one of the men to cut the garden the last time that it was planted? When planting began, the primary rights holder apologized to his female relative and her husband, explaining that he saw that the garden was ready and began cutting with the intention of inviting those with rights to join later in the planting. In the end, they both planted large blocks within this new yam garden.

One's affiliation with a particular group does not absolutely exclude hunting and use rights in other areas. In fact, maintaining linkages to multiple areas is important in order to maintain the "road" for one's children. Because parents retain the right to designate their children in any of their parents' or grandparents' affiliations, they need to retain good relationships with their siblings as well as more distant cousins (brothers and male cousins are referred to as *sakpek*, and sisters and female cousins are *sakmere*). Not doing so could cut off possible roads for their children. In the case above, the family wanted to ensure that their son would have access to this area. While they were planting the garden, they indicated that it was planted for their son. When a garden is planted in a child's name, it links the child with the use of this area. Even if his parents held only secondary rights, the child, male or female, might someday gain primary rights. But certainly the child should be invited to participate in future planting. In 2015, both families returned to plant this same yam garden again.

As a result of marriage, urban migration, migration between villages, adoption, and the multiple rights that individuals can claim, resource rights can be quite distinct from residence. However, residence does place certain limits on the networks of relatives and the level of commitment (M. Strathern 1996). For example, when I asked a Winima man to name all of the coffee gardens claimed by him and his wife, he first listed three before settling on two. His wife objected to including the third garden, which was located some seven kilometers away in the village of Wandumi. This garden had been planted by her father on land he gained access to through his second wife. Though it had been given to her, it was now "looked after" by her father's second wife's family. She had given up this land when she was sent to her father's ground in Winima. In configuring her own placepersonhood, she chose to "cut off" lands that were within her right to claim, while her husband wished to extend them to include areas that might be claimed by their children. However, the network was never completely severed, and one of their children would reclaim these lands after marrying into Wandumi many years after our interview.

Hunting complicates this slightly and serves to demonstrate the highly contextual nature of land rights. A male gains access to the routes and hunting lodges associated with each of his *solorik* affiliations, regardless of primary or secondary designation. The only exception to this rule is during disputes, when the network of rights is not so much cut, but contact is avoided. Women, who do not participate in hunting, do not actively engage rights held by their brothers. However, a woman's male children will be able to follow hunting paths and visit lodges with her male relatives, and husbands can be invited to follow these paths as well. Thus at first glance, hunting rights suggest a patrilineal bias, with only men activating rights in the routes and paths of the forests. Women serve as intermediaries between fathers and sons. This system is limited to hunting, while gardens are imagined as more egalitarian spaces. As the mine and conservation areas rest on hunting lands, men have tended to claim greater authority over the projects. As a result, there is a tension between the activation of rights by men through hunting and women's possession of potential rights on behalf of their own male heirs. Ultimately, both have rights but are limited in their ability to activate them.

GATHERING THE LAND THROUGH MARRIAGE

In 2001, I observed a marriage ceremony in the village of Werewere, during which a yam was introduced into an otherwise Protestant service. At the end of the ceremony, a senior Biangai man presented the couple with a cake pan in which a cooked yam had been prepared. It was announced that the Biangai had "borrowed many things from the white man," but this is Papua New Guinea, where yams are more common than wheat. Instead of cutting cake, they would cut the yam that her family had prepared. Moreover, he said, for the Biangai "the yam is our bones": it had been planted by generations of ancestors up through the couple's parents. The analogy continued likening the yam to "our liver" (akin to heart), just as one's spouse is "one's liver." In pre-mission marriages, yams were cooked and served by the bride and her mother for the husband (Ouellette 1987). Otherwise, the exchange of wealth tended toward equality. Biangai strongly emphasize that for them women have always been "free" and are not viewed as circulating in association with a bride price. Instead of joining a lineage, the couple is ideally seen as forming a distinct household separated from the production of the larger relations, though obligations persist on both sides. The yam in this wedding service linked the couple to the land and signified the couple's rights to her resources as well as his.

Marriage brings together not only two families but also two sets of gardening and hunting lands. The marriage practice of *koronga koyange* (from daughter to son), commonly called *marit long graun* (married through place) in Tok Pisin, highlights the significance of *solorik* flexibility in the group affiliations. Such flexibility allows families to create locally productive matches. Through this principle, the designation of land rights inherently established the appropriate set of marriage partners. Such marriages are valued because they *bungim graun*, or "gather the land." As Ouellette points out, "Marriage is imagined as a reunion of parcels of land through respective men and women who have rights in the same land" (1987, 105; my translation). Historically, through a preference for marrying third-generation cross cousins, the primary rights of a husband and wife ideally reunited parcels of land separated by previous generations (figure 2.1). The marrying couple ideally shares a common ancestor at least three generations back. Parcels of land accompany each new generation, so that ultimately the couple's rights recombine those of their last common ancestors. Neither women nor

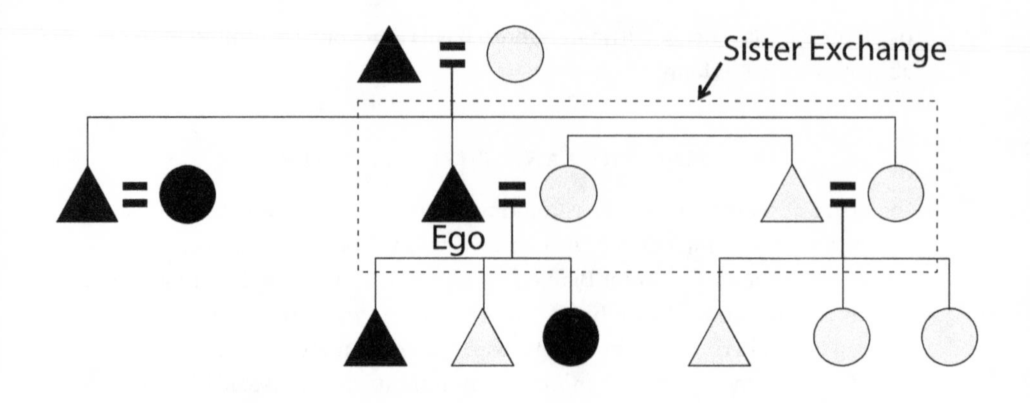

FIGURE 2.1 Kinship chart illustrating "gathering the land" through marriage

men circulate in this system. Instead, it is conceptualized as the movement of persons through land.

Occasionally, these parcels were on the borders between existing *solonarik* groups, providing an opportunity to unite two other groups. This was done preferably through "sister exchange" (just as easily thought of as "brother exchange"). The marriages were considered suitable because the parent's generation of both ego's (the focal individual in figure 2.1) and his wife's siblings gather the land between Yaumura and Kaiwi. They continue this practice in their respective marriages. In particular, ego's older sister retained rights to her father's *solonarik*, while the younger sister was placed in her mother's. Ego married the older sister of his younger sister's husband. Their children also reflect these divisions. At the time of my interviews in 2001, ego wanted to place one of his children in Yaumura, but his brother-in-law, angry over land rights around the Kuper Range Conservation Area, refused to admit that such arrangements existed for his children. After the conservation area fell apart, objections subsided. The example demonstrates the possibility of claims to multiple *soloriks* and the role of resource projects in shaping the contours of places and persons.

Today, gathering the land is more commonly done in hindsight, after the marriage has been planned. Youth prefer to "marry according to like," which their parents fear will decrease the size of their individual allotments and the allotments of future generations, as well as dispersing their land rights across the landscape. This is said to result in many disputes, as individuals planting on smaller plots may argue with their neighbors about boundaries, or with their spouses as they travel greater distances to different gardens. In the past, distances between gardens also put the couple in danger of attack from Biangai enemies or even other Biangai villages. In spite of such concerns, Biangai increasingly wait until marriage before assigning primary rights, gathering the land in hindsight so as to ensure that such problems do not arise.

KLENS: DEVELOPMENT PROJECTS AND THE CREATION OF LARGER AFFILIATIONS

While *solorik* groupings still retain prominence in village politics and the organization of resources, new affiliations increasingly play a role. These are commonly referred to as *klens* (clans). With the advent of logging, mining, conservation areas, road construction, and other activities necessitating compensation, Biangai communities have been encouraged to identify clan groups in line with government and corporate expectations (Mitio 1981, 1984; Ouellette 1987). Such groups are perhaps best thought of as the solidification, or entification (Ernst 1999), of *solonarik*-based groups, initially organized around the same principles and associated with a given area. Because the names of these groups retain authority in discussions of land use and compensation from external agents, they are increasingly fixed under the Tok Pisin name *klen*, a product of regular government registration for development efforts (i.e., logging and mining) and census taking, where "clan" affiliation is asked.

The novelty of the concept has resulted in varied definitions. One Winima man explained that members of a *klen* come from the same historical villages. *Klen* affiliation, then, was centered on places and not mythical ancestors. However, when asked if one of the groups, such as Paro, was a *klen* or a *solonarik*, responses were mixed. Many initially said that it is the same but then qualified that it was either too big an affiliation to be a *klen* or somehow not flexible enough to be considered a *klen*. Thus, *klens* lack the clearly distinctive traits associated with "clan-based" systems and do not have a consistent meaning when applied locally. However, they express contemporary idioms through which relations among persons and landscapes are organized for development. Here, I use the term *klen* to mark the categorical distinction between a unilineal system of descent from a stipulated ancestor—a clan—and the processes associated with the formation of these Biangai groupings.

Often, government, corporate, and NGO agents have assumed that these *klen* affiliations retain the same qualities as the English term "clan" implies. For example, the Wau Ecology Institute's director, Harry Sakulas, celebrates the identification of Biangai clans as a successful aspect of the organization's conservation efforts (Sakulas 1998). Among communities, claiming the broader *klen* category became an important strategy for ensuring that rights are retained to resources, as compensation was linked to specific membership. Thus, tensions often emerge between *klen-* and *solonarik*-based groups, as individuals seek to improve the lives of their family, gaining compensation and garden lands for themselves and future generations. As geographer John Burton noted in his social mapping for Hidden Valley Gold Mine, "The solorik system is vulnerable here, because if the rights are unevenly spread, villagers may attempt to 'jump into' the two lead soloriks to get access to cash. This is highly undesirable because the soloriks are needed to function properly for other reasons; distorting them will upset village affairs as a whole" (1996a, 7). In spite of Burton's warning, the mine has done little to alleviate these concerns. Fortunately, as the *solorik* system allows for flexibility and blurred boundaries, Biangai can reposition themselves to maximize their access to certain lands. During the 2000 Papua New Guinea government census, many joked about what *klen* they would claim to position themselves advantageously. As we sat with the census officials on the porch of a trade store, Biangai complained that only one *klen* could be listed, though they self-identify with several at a time, and a few announced that they might just switch *klens*. One young man proclaimed himself to be Kaipa, recognizing his connections to both Kaigowe and Paro, and merging the two into an increasingly common name among youth. Some in Winima privately admitted to going against the wishes of their parents, choosing Igulu group names that represented a secondary affiliation. By doing so, they repositioned themselves in the compensation disputes and better reflected personal allegiances.

LOCAL REPRESENTATIONS OF KINSHIP

In response to logging, mining, and conservation projects, Biangai started to keep written records of their place-based relations. These aid in tracing location-specific *solorik* affiliations. However, they have developed a unique representational style that addresses both the external demands for *klen* designation and Biangai ideas of relatedness. In contrast to the standard ethnographic system illustrated in figure 2.2, Biangai representational styles highlight the emergence of multiple lines and not necessarily generational and gendered distinctions. For example, figure 2.3, derived from one page in a collection of charts kept by one household, illustrates the same set of place-based relationships highlighted in figure 2.2. In general, the row of boxes to the far left is the first line from which many stalks (*solorik*) are derived. Siblings are ordered, and do follow a line away from the parent but are further organized by named kinship group. Only those children who can claim membership in this *solonarik* are included. Furthermore, the boxes contain not merely a single individual (as standard genealogical representations entail) but also the spouse, suggesting a unity that is lacking in the genealogical methods of Rivers (1900, 1910). In many ways this matches the argument that "being 'male' or being 'female' emerges as a holistic unitary state under particular circumstances" (M. Strathern 1990, 14). In this case, the category male or female is less important than the places that a couple unites or how they attend to *ngaibilak* relationships. For the Biangai such representations speak to the ways that marriage recombines places and bundles of relations through the household.

The relationship shown in figure 2.2 is filled with numerous illustrations of kinship relationships, each tracing ancestry through multiple *solorik*-based affiliations. However, this standard nomenclature differs from the notebooks that Biangai maintain. Each page in their notebooks represents a different reading of the same ancestry, often including the same people but different aspects or parts of a person's relationship to place. Biangai collect and maintain such records throughout both Winima and Elauru. This Biangai representational system is of interest for what it reveals about the value of standard genealogical nomenclature in light of local models of relatedness. The usefulness of the former for sorting relationships in abstraction is well worn, but the Biangai representational scheme illustrates central concerns and dynamics within their system of relations.

Of interest in this case, among the individuals represented in figure 2.2, those alphabetically labeled are involved in a yam garden cut by A and his son, B. While B is assigned to Yaumura, he retains secondary rights in Yaru through his father as well. They are labeled according to the land group in which they claim primary membership, though this label ignores their secondary rights. Much of the land used by A is derived from his mother's Yaru *solonarik* (at the time of my initial census thirteen coffee and food gardens were on his mother's land, only two on his father's land). His wife lacks access to such resources and predominately relies on her husband. She is from another village and was sent to "activate" rights in Kaiwi, gathering the land

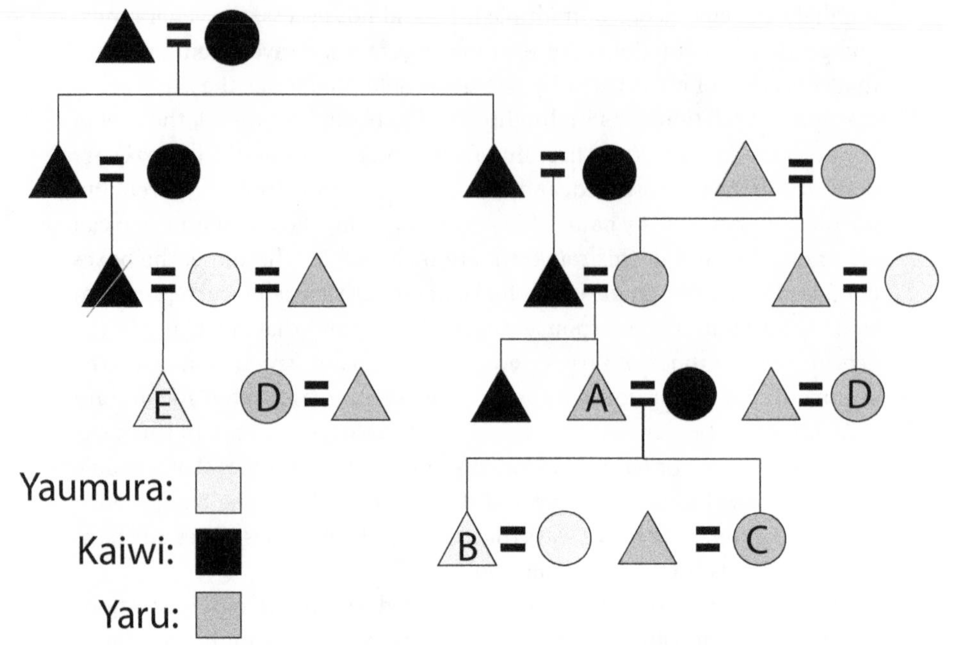

FIGURE 2.2 Kinship chart demonstrating standard genealogical relations. E was adopted by D's father, giving him multiple rights and multiple connections to the Yaru yam garden in question. D is listed twice, as her rights intersect at two separate points.

of an ancestress with those of A. The opposite is true of C and D, women who married men with primary rights elsewhere. In the case of C, her husband is from a village outside of Lae, while D's husband was adopted into his current affiliations under the Yaru name. Their marriage acknowledges his claim on Yaru lands. However, this area is also associated with C, D, and E, each of which trace ancestry to an apical figure from whom their Yaru rights are derived. E presents a special case. In the kinship diagram, he is clearly related to A through A's matriline and patriline. He was born into the latter but adopted into the former after his father's early death. In this case, his rights through his adopted father are being acknowledged in the planting of the garden.

In contrast to the standard genealogical system, Biangai representations (see figure 2.3) focus on a single *solonarik* pattern at a time. Following the listing of rights made during speeches about land disputes, they trace only the significant aspect of the relations involved. In establishing claims to this

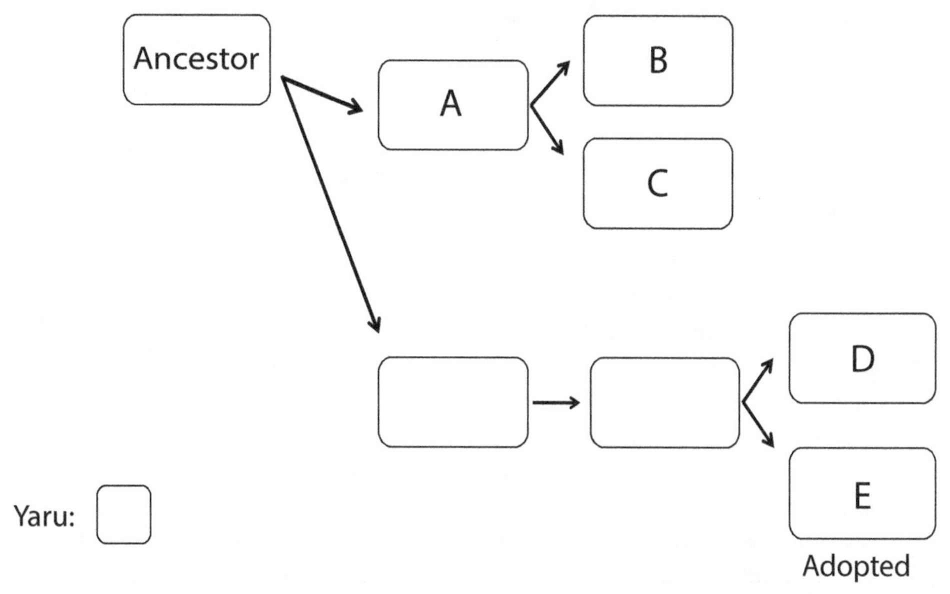

FIGURE 2.3 Biangai representational system drawn from the relationships in figure 2.2. Each box would include the names of a married couple. As it traces both primary and secondary rights, this representational system deemphasizes differences in order to show shared relationships to place.

particular garden, Yaumura and Kaiwi claims associated with the household are ignored.

Following the model used in figure 2.2, figure 2.3 is a Biangai representational system focused on the Yaru landholding group as a defining feature for determining relatedness. The individuals included are the same as those in the standard genealogical representation, excluding those that lack rights to the yam garden in question. Here the relationships among couples is represented not in terms of their multiple and varied positions within landholding groups, but in terms of how they find their way from the base ancestral placeperson to the present situation. This includes B and E, whose primary rights are in a different *solonarik* and whose use of the yam garden in question is through secondary rights. Place, rather than blood or genetic relatedness, defines affinity. Therefore, "under this particular circumstance" (M. Strathern 1990, 14), being Yaru matters more than being Kaiwi. These representational practices do not simply follow the matrilines and patrilines but detail the complex path of claiming places through different ancestors by engaging relationships among placepersons. One might follow one's father's mother's mother, while another might follow her mother's father's father to the same parcel, and yet another might be adopted. *Solorik* groupings are as much, in the end, about the places as the kin.

• • •

Biangai sociality is reproduced in their relationship with places. Places and persons mutually inform one another, defining place through sociality and sociality through place. Historically, places were sites of great events, marking social relations through gardens, hunting paths, residence sites, mythohistorical battles, and warfare with neighboring linguistic groups. But these places were always already invested with genealogical meaning, not only making the events formative of the *solonarik* groupings, but intimately tying place and persons together over time. Biangai kinship is truly grounded in place, made through gardens and *solonarik* affiliations. As such, changes in one shape the expression of the other. For example, colonial and postcolonial gold mining places pressures on how Biangai relate to their lands, suggesting a reconfiguration of this relationship through *klen* affiliations. But the *solorik* allows for Biangai to be responsive to these changes, molding social groups around novel contexts and maintaining their relationships through places while their government and company emphasize static clan affiliations.

The relationship among persons is at stake in different categories of work and the resulting sociality because of changes in the values of certain places. Working the land, working for conservation, and working for the mine highlight these relationships but also reveal how Biangai human-environment relations inform their response to these differing regimes. Their ongoing commitment to place provides respite from what might otherwise be devastating transformations of community social relations.

CHAPTER THREE

Working the Land

BIANGAI ideals and practices associated with kinship and access to land are being reimagined in light of current development efforts. Place is central to sociality and social reproduction, but government- and resource-based neoliberal projects seek more bounded social groupings. Agricultural work is a way in which Biangai create and maintain relationships with place, and work, as an embodied practice; it is not simply a mode of production but also an engagement between gardener and the physical landscape.[1] Through working the land, Biangai manage the spaces between kin as a "space that relates" (Wendt 1999, 402) and "sustains one's life" (Wood and Winduo 2006, 85). While conservation practices irregularly interrupted daily routines for short periods, mining labor can be transformative of everyday life. How are practices of gardening being reshaped by the experiences of mining?

The ethnography of work in Papua New Guinea is highly varied, ranging in focus from subsistence activities, colonial plantations, and village-based cash crops to recent considerations of laborers in corporate activities such as logging, mining, and sugar refineries.[2] Generally it is acknowledged that changes in subsistence practices can effect changes in land tenure or social relations. Different dynamics are involved in yam and other subsistence gardens and the relatively recent introduction of coffee cash crops. The latter is the most demonstrative of ecological collisions, highlighting how Biangai reimagine land and relations through place. But yams and other subsistence crops provide a dialogic image of Biangai sociality and connections to place, revealing how garden work continues to ground social life and provide resistance to the forces of neoliberalization by keeping them in place.

In literature, especially of a romantic bent, working the land is celebrated as bringing one closer to "nature" (e.g., Thoreau 1854). However, it is a connection imagined as a return to a celebrated past. That which is called nature, the land, or the environment is seen as separate, distinct from the person who must reconnect. For example, Wendell Berry's book *The Hidden Wound* ([1970] 1983) details how slavery and then racism served to sever ties and values associated with the land and its work. For Berry, this division of labor, which one finds in Western societies (and those that they colonized), is divided between those who work the soil and those who manage it. Writing about his grandfather and an African American in his grandfather's employ in Kentucky, he notes:

> When my grandfather went to the field his mind was burdened; when Nick went to the field his mind was free. The difference can be illustrated by imagining two figures in a landscape, one of them trying to determine how that landscape can be made to produce the money necessary for the next year's interest, and the other conscious of the wherabouts of the dens of foxes, planning a hunt. And the knowledge I received from those two men is divided in exactly the same way—the two halves, you could say, of a whole relationship to the earth. (Berry [1970] 1983, 18)

The two men differed in how they thought about labor, and their relative concern for the land follows suit. One sees land as a commodity, the other as a part of his life. Berry's romanticized image is not without its problems, but it does offer insight into the complex ways that Western discourse separates humans from place. An obvious reconceptualization of the domination of nature thesis proposed by Frankfurt School theorists (e.g., Horkheimer and Adorno 1972), Berry's work is an attempt to reconnect the two modes of relating to the environment. Wilderness, fields, forests, and even croplands are separated in industrial nations from the urban. Through a return to working the land, Berry suggests that such connections can be remade—but more interestingly, that working the land can be about connecting to the land and managing it "for next year's rent" as well as viewing it as part of one's self. Like Biangai, work becomes a way of knowing and inhabiting nature as well as providing for "next year's rent." However, Biangai approach this not as a contradiction that needs to be remade but as an ongoing potentiality of their world-making practices.

Gardening is not only a mode of production but also an engagement among gardeners, the physical landscape, and cultivars—in this case, yams and sweet potatoes. Gardening speaks to skills associated with dwelling in an environment (Ingold 2002) and moments of innovation as gardeners work with the soil, seed crops, and the physical spaces of land. But it also reveals moments of "becoming with" (Deleuze and Guattari [1980] 1987) when the gardener is cultivated by the products and places he or she touches. The land tills the mind of the gardener, the seed grows into food that nourishes the body, and the work assembles a multispecies genealogy (Haraway 2008; see also Roberts et al. 2004). In turn, the family grows with the space, demarcating land and circulating the seeds of plants. Haraway's (2008) questions about her relationship with her dog evoke these sorts of relationships. Through training, practice, and care they learn to play and work together, becoming part of each other's lives. Likewise, gardeners, land, and cultivars mutually invite one another into meaningful ecological relationships as equally emplaced persons and things. Ideally, gardening is about the "support and caring for relationships rather than the pursuit of wealth for wealth's sake" (Lilomaiava-Doktor 2009, 20), though this is being challenged along the frontiers of capital.

Through Biangai gardens we can explore the ways in which these multispecies encounters are shaped by other ecologies, other ways of knowing and working the land. Mining can result in "colliding ecologies," where downstream damage from mining wastes devastates the subsistence base of riverine communities and forces a revaluation of human-environment relations, and can "disrupt their relationships to places including individual memories and social histories" (Kirsch 2014, 51). While there is evidence of such collisions along the Watut River side of the mine, which is downstream from the tailings dam and water wastes from the Hidden Valley Gold Mine, Biangai don't face such stark ecological forces. The Upper Bulolo River, which runs through their villages, is not filled with silt and stone from mining. Their rivers and streams still run clean. Instead, they are confronted with *colliding ecological relations*—practices and social relationships that shape how they know the land and how they know each other through place. Colliding ecological relations are as much a product of competing regimes of work (including employment at the mine) as they are a reflection of the explicit attempts to transform subjectivities through the mine's programmatic emphasis on corporate social responsibility (see also Welker 2014, 129–56). Designed to appeal as much to investors as local stakeholders, corporate social responsibility acts as a kind of neoliberal governmentality—especially in the absence of a strong state—to elucidate new subjects that suit corporate needs. However, local practices and individuals are not "easily" transformed "into autonomous, responsibilized neoliberal subjects" (Welker 2014, 154). Instead, while new possibilities for individual subjectivities are realized, most Biangai remain very much committed to their own ecological ideals. Comparison of the gardening practices of Winima villagers, who are landowners and beneficiaries of the Hidden Valley Gold Mine, with neighboring Elauru villagers, who are not direct beneficiaries, reveals emerging complexities that are realized in the colliding ecological and social relations supporting mining in Papua New Guinea.

For Biangai, yams are propagated from other yams, planted in gardens once planted by one's father and mother, grandfathers and grandmothers. The soil that was turned and planted before is returned to and planted again. These repetitions are not exact duplicates of the past but rather actualizations of new assemblages (Deleuze and Guattari [1980] 1987). The past appears in the landscape of the garden as ghost (Halvaksz 2008a), as history (Halvaksz 2008b), and as genealogy. Such appearances represent the physical and aesthetic qualities of a landscape that are created in the folding and turning of the soil by groups entangled in commitments of kinship but also in the work of gold mining. The scattered plots that once attracted gardeners to the soil are revalued in light of veins of gold, ripening coffee trees, and other cash crops. At the same time identities are essentialized around the land,[3] and new entities are formed around kinship (e.g., Ernst 1999; Gilberthorpe 2013).[4] In this process of reshaping relationships, placepersonhood remains a source of hope.

WORKING IN BIANGAI GARDENS

Organized around *solonarik* rights, all gardens are a realization of Biangai sociality through the land. However, not all lands are equal, as each community has access to grass-covered slopes and secondary forests of varying quality. Biangai distinguish three zones for planting: secondary forest (*yengelei ngezeg*), grassland (*yengelik*), and village (*yerenda*). Forested lands are favored for all crops, though resource degradation and population demands have increased the reliance on grasslands. The lack of time to travel long distances to forest gardens further complicates land use, as corporate work schedules, coffee gardens, and other commitments favor gardening in close proximity to the village, where grasslands predominate. In Winima, many of the gardens are within the boundaries of the village itself.

While Biangai plant and tend to a variety of crops, three categories are of interest here: yam gardens, sweet potato/mixed gardens, and coffee gardens. For each there is a conceptual or historical relationship between Biangai villagers and the practices associated with the crop, labor involvement, and specific practices in Elauru and Winima. Social relationships associated with yams and sweet potatoes as well as longitudinal data on these crops, coffee cultivation, and Biangai distinctions between gardening and farming are likewise relevant. A growing body of literature that informed my initial research suggested mining negatively impacted community agriculture, while conservation was celebrated for its less-invasive "fit" with community life. It seemed reasonable to expect that Winima would reflect such changes (Halvaksz 2005), but research has revealed a much more complicated picture.

BECOMING WITH YAMS

Biangai grow a variety of yams, primarily from the species *Dioscorea alata*, *D. esculenta*, and *D. rotundata*. Yams and yam gardens take on great symbolic importance for Biangai men and women, with many named varieties proliferating from the three types identified in Western science. Like the matsutake mushrooms discussed by Anna Tsing, the yam "is not just a delicious food; it is also a valued participant in a world of ecological well-being" (Tsing 2015, 15). For Biangai, the yam is a highly valued sustenance, a metaphor of gendered human interactions, and it is intimately linked to gardeners through mytho-historical genealogies. The work that gardeners perform in the production of yams parallels the reproduction of children, with gendered responsibilities rhetorically and physically mimicking conception.

Yams are planted when a constellation of stars known as *ruwere* are spotted in the sky (the constellation Pleiades). *Ruwere*, which was translated for me as "stars gathering," is closely associated with *ruwerek*, referring to the cordyline planted in gardens to mark boundaries. At least one informant mentioned to me that it might also be a derivative of *ruwerik*, a *solonarik* group name that has fallen into disuse, meaning *graun ibruk* (in Tok Pisin) or "landslide." Yansom described how his ancestors would see these stars gather on the horizon at dusk and raise a knife or an axe toward the sky while calling out the name of an area. The next day a family group would go and cut a garden there. But he went on to lament that "we no longer do this, so that is why our yams are no longer good." He was referring to both the size and quality of the yams, as well as youthful memories of grander times (e.g., Clark 1999).

Having identified a suitable yam garden, one that had lain fallow for some fifteen to twenty years, a group of men cleared the forest. As fallow periods have decreased, these old gardens often don't return to secondary forestlands, and men complain that the land thus lacks *gris* ("fat" in Tok Pisin). There is some disagreement on how to discuss the same substance in the Biangai vernacular. While some follow a literal translation, defining *gris* as *koloko*, others explain that *koloko* only refers to fat that makes food taste good. Instead, rich soils are described as cooked (*ngaibilak kuk*) or hot (*yezi*).

FIGURE 3.1 James breaks up the soil, preparing the ground for the yam.

It is made hot by the debris of the understory compost (*ngaibilak puk*). Cutting and burning the secondary forest increases the heat. If the area has completely reverted to grassland, a controlled fire is set to clear it; otherwise the debris is left to dry for a few weeks before burning. Grasslands lack the *heat* of the forest garden and are not seen as productive.

Men use long, sharpened sticks to loosen the ground, turning the soil and using their hands to clean it of stones and roots. The turned earth expands as the pore spaces in between particles of soil allow for the circulation of air, water, and microorganisms essential to growth. Decompacted, the soil overflows the original hole. From this, mounds are formed, in which a yam is carefully inserted with protective spells. Then, women tend to the gardens, cleaning and heating the surface with fires that discourage insects and blacken the leaves with smoke as they find their way up the sticks. Yams are likened to a baby and must be carefully nurtured by the women as they develop. The innuendo is intentional, as yams parallel human genealogies. The very first yam is said to have grown from the body of the first man. As yams are propagated from seed yams and shared among family over generations, yams consumed by gardeners today are directly related to those planted by the earliest ancestors, with both yams and humans sharing descent from the first man. Yams form part of the "network of attachments and connections beyond the human" (Besky and Padwe 2016, 13), but only

FIGURE 3.2 James planting a yam. After the yam is removed from a larger tuber, the exposed flesh is bandaged with banana leaf, demonstrating the care given to the crop.

because they are humanlike. The yams themselves share names with parts of the body, further attaching yam selves to human selves (e.g., Kohn 2013, 75).

In planting, yam and gardener mutually inform one another as they tend to what philosopher Donna Haraway calls "the dance of relating" (2008, 25) and Fijian educator Unaisi Nabobo-Baba (2006) might simply call love and support. The vines of yams, for example, lack the tendrils seen in beans and other twining plants, and in the wild they must find their way through secondary growth. However, in yams' dance with gardeners, sticks are placed at the edge of each mound, by which men and women carefully ensure that the vine finds and stays with its partner. The seed yam, having depleted its energy in the initial stem and root development, withers, and the growing leaf system eventually "accumulates carbohydrates in excess" (Lebot 2009, 230), stimulating tuber growth. Setting fires to blacken the leaves fosters further leaf development, increasing the size of the tuber in the process. The gardener's care is directly related to the growth and eventual size of the yam, as he provides sticks of sufficient length for twining, mounds and cleans the soil to remove impediments, and tends to the surfaces with fire and smoke to protect the tuberous flesh from insects and disease.

Yam garden labor is also highly gendered, with men conceptualized as giving strength to the yam in the same way that they give strength to a child.

They cut the secondary forest, opening up space for the garden to grow (or, increasingly, cut the kunai grasses closer to the village). The dried wood and undergrowth are burned, rejuvenating the soil and removing harmful insects and plants. Over the course of several days, men dig and clean the holes in which yams are planted, filtering out all debris. "The yam will grow to fill the space," I was repeatedly told. They provide the many two- to three-meter sticks around which the young vines twine. Women tend to the yams, caring for them in the belly of the garden, cleaning and weeding regularly as they accumulate carbohydrates. At harvest, once the leaves have started to dry, women dig the yams out of the ground, being careful to remove them in one piece. A broken yam is a disappointment. Men and women speak of their parents planting yams that grew a meter or more in length. Now, they complain, yams don't grow as well, a statement that echoes concerns for the land's well-being and their own moral relationship—as men and women—to the past (Clark 1999; Lipset 2017).

Once the yam vines have reached the ends of the sticks, Biangai begin to plan for their harvest. I have witnessed five yam harvests over the years, and each time the yams were first dug in association with a funerary feast—either the initial feast for the work of burial or the final feast marking the end of a yearlong mourning period, when cement markers and engraved plaques detailing the person's life are placed in the cemetery. During these events, those who supported the mourners throughout the year and contributed to the costs of the burial rites receive a return gift of yams, sweet potatoes, pork, lamb flaps, and other foods. Following this initial harvest, the yams are dug as needed for food and exchange. Few, if any, make it to market.

The gardens themselves solidify relationships with the living and dead, as they too follow genealogies. Seed yams are passed among family, as both yam varietals and Biangai speakers descended from the first man. Men and women inherit the right to plant through their parents, rights that are solidified through the experience of planting with them and then on their own. While a single yam garden could be divided among members of a single household, or even an entire village, individuals do not restrict their labor to their individual plots. Instead, men help a brother or more distant relatives, which they say makes the work go faster. When it comes to the actual planting, a senior man will often help place a yam in each plot, using magic to protect the crop. During the time of their ancestors, this allowed for more successful gardeners, possessing magical knowledge, to ensure the success of their cogardeners. Magic was also used to ensure that yams in rival gardens did not grow. Now, such practices are almost lost.

Consuming yams is likened to a form of consubstantiation, connecting Biangai gardeners with ancestral agencies through both the flesh of the yam and the gardens in which they are planted. Yams and yam gardens grow the body and the group (O'Hanlon and Frankland 2003), adding to the skin and appearances of their gardeners (Bashkow 2006). But they also mark the ideal ecological relationship among Biangai, place, plants, and animals: one of interdependence and responsibility.

GARDENING WITH SWEET POTATOES AND OTHER PRODUCE

Contrary to common perceptions in the United States (where the names are used interchangeably), yams and sweet potatoes belong to different orders of plants (*Dioscoreales* and *Solanales*, respectively). Both grow in the Bulolo Valley, but for Biangai, yams are more important than sweet potatoes. While they grow a variety of species, sweet potatoes are relatively new (Bourke and Harwood 2009) and are valued for pig production, resistance to frost, and year-round growth. Yams rely on changes in the amount of sunlight over the course of the year to stimulate different phases of its growth cycle (Lebot 2009) and can't be planted at certain altitudes or year round. Sweet potatoes have a wider ecological range (Bourke and Harwood 2009) and are also viewed as easier to grow. "You just plant them or the vine and they come up," I was told. In contrast to yams, sweet potatoes don't need sticks to twine, running along the ground like other potatoes. While yams taste better, historically the bulk of caloric intake comes from the sweet potato. Today, rice has taken on this role in some households (reflected in their labor efforts), also marking the growing inequalities caused by mining.

Initially, many sweet potato gardens consist of a combination of indigenous and introduced foods that are planted together in multistoried gardens. Like the yam, men do the initial clearing in large work groups "grounded" in the *solonarik*. They set fire to the area and remove any large debris. Once this task is accomplished, both men and women plant throughout the garden, though women tend to do most of this planting. While some gardens are cut for just sweet potatoes, other gardens are originally planted in yams and then replanted with other crops as the yam season nears its end. Those who lead the yam garden retain the rights to plant subsequent gardens. As shown in map 3.1 of a small area planted in recent years, gardens can be utilized over several seasons as they transition from yams to mixed crops to pure sweet potatoes. The pattern of shifting to a nearby location is also

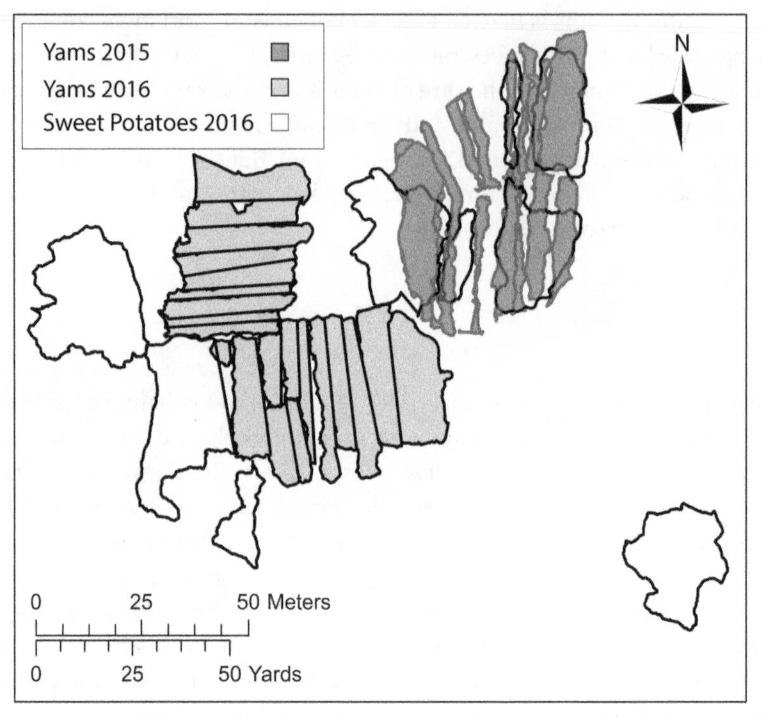

MAP 3.1 A family garden area is returned to over the course of four seasons. Each mapped area is planted by a family member related to the group through the garden.

apparent, as the 2016 gardens prepared for yams by these families were in close proximity to gardens from the previous year. Each household plants a long row of yams adjacent to their relatives' rows, with the largest garden typically taken by the primary rights holder. Sweet potatoes are much more expansive and don't reflect such sociality. In this case, only three principal families retained rights to plant sweet potatoes in the old yam garden.

While they plant a variety of indigenous bananas (both sweet and cooking), sweet potatoes, taro, cassava, and greens, they also plant many recently introduced crops, including beans, cabbage, corn, tomatoes, potatoes, peppers (both sweet and hot), onions, green onions, watermelon, pumpkins, and carrots. Avocados, papaya, mangoes, oranges, and lemons are also quite common tree crops planted throughout the village and coffee gardens. As with yams, much of this food is for consumption in the household; however, a portion finds its way into both village and Wau markets. As these foods are harvested, sweet potato vines slowly spread across the garden.

Many of the Western crops were introduced during the early stages of colonial mining, with Biangai communities providing numerous items to both small-scale and company miners early on. Some of the older men spoke of selling vegetables door-to-door in town or helping their mothers carry items to the Wau market. In the late 1970s and early 1980s, a government corporation called the Fresh Food Market bought produce from throughout the valley and sold it again to hotels and stores in Lae and Port Moresby. Coming before coffee really took off in the villages, the Fresh Food Market provided a modest income to local farmers. Françoise-Romaine Ouellette reported that during her fieldwork (1976–78) Elauru made 1,008.61 kina per annum, while Winima earned 988.74 kina per annum, with the last year producing almost twice the average (Ouellette 1987, 234). By the time the Fresh Food Market ended in the early 1980s, it had established significant variety in local gardens.

During the late exploration phase of the Hidden Valley Gold Mine, Winima (and Kwembu) had the added option of selling produce to Morobe Goldfields on a weekly basis for use at the mining camp. This did not amount to much, as a small exploration crew staffed the camp. However, Rose, the daughter of the village councilor, organized these purchases once a week on contract from the company. She bought enough food to meet the company quota and encouraged community members to bring in needed items. This motivated increased specialization in gardening, as each family tried to meet the demands of the company. For example, on one collection day, many women arrived with an abundant supply of lettuce, much more than the company would need. Rose chastised those bringing lettuce for not communicating beforehand and for not planning ahead. Such planning and organization reterritorializes the garden as a site of exchange value, as participants organize production to meet the demands of mining. Today, Winima hosts a Saturday market that does much of the same work as the company's purchases: training people to think about producing a surplus so that they might sell it, as opposed to producing for consumption and exchange. Market participants are careful to avoid competing with one another by developing known specialties or opting to purchase goods in town for resale.

COMPARING GARDENS

Given its importance to the social life of the community, comparing garden production in Elauru and Winima quickly became a significant aspect of my research as a way to gauge the relative impact of conservation and mineral development on community life. Starting with my initial fieldwork, my research assistants and I counted yam mounds and coffee trees, measured the area, and surveyed each garden plot. In 2001, using a Garmin GPS and a tape measure, area was calculated with a simple length and width measurement and plotted with a single point. Though the area estimates are subject to a large margin of error, the hand counting of yams and coffee did demonstrate great differences in the two communities (Halvaksz 2005). The results, however, seemed too predictable (with mining having a greater impact on Winima than conservation on Elauru), and there was no way to account for annual variation, weather, varied labor opportunities, and other factors. Was 2001 just an atypical year? This necessitated subsequent research to obtain multiyear data, using more precise instruments to calculate area.[5]

Beginning during the later stages of the conservation project in Elauru and the exploration phase of Hidden Valley Gold Mine in Winima, and spanning the demise of the conservation effort and full production of the mine, the garden surveys highlight a dynamic transformation of labor effort and agricultural production. While there are years that seem to negatively reflect the impact of the mine on community life, there is no single trend (table 3.1). Instead, the data reveals the intersection of labor opportunities, socioeconomic relationships with the mine, El Niño climatic events, internal disputes, changing relationships with migrant labor, and broader ecological differences between Winima and Elauru. For example, yam production in Elauru declined in 2015 with the death of a senior man who was significant for his knowledge of yam magic. His death filled the gardens with sorrow, one man explained, as his *imeng* (spiritual presence) hung over the fields. The years 2015 and 2016 reflect both the drought that was associated with El Niño and declining production at Hidden Valley (compare the years shown in map 3.2).[6] While the drought proved detrimental in Elauru (decreasing the area planted and forcing the gardens farther from the community; see map 3.2, 2016), the rich soils of Winima provided some respite for its residents. By contrast, the drought had a positive impact on Winima's overall production; mine employees laid off due to water shortages returned to Winima to plant gardens once again in 2016. Throughout these variations in garden effort, what is of interest is the ongoing cultivation of yams. Yams remain an important aspect of Biangai identity and practices, connecting them to places and to ancestral relationships. In contrast to my initial findings, yams proved to be resilient, affecting ongoing connections to place.

One distinction that is apparent involves the differences in garden clustering. Comparing the dispersal of gardens on the left (Winima) of the maps with those on the right (Elauru), distinctions in the proximity of families

TABLE 3.1 Comparison of Elauru and Winima yam and subsistence gardens

		2001	2011	2014	2015	2016
Elauru	**Population**	235	291	271	274	278
	Yams/person	7.67	4.67	3.93	5.57	7.51
	Garden area/ person (m²)	199.69	133.21	160.75	104.75	80.90
	Total area (m²)	46,927	38,764	43,563	28,702	22,490
Winima	**Population**	238	329	212	200	188
	Yams/person	4.11	4.78	6.38	8.20	8.38
	Garden area/ person (m²)	129.2	49.65	190.25	143.20	160.96
	Total area (m²)	30,750	16,335	40,333	28,640	30,260

are apparent. While Elauru tend to cluster near one another, Winima gardens are quite dispersed. In part, this reflects the periodic nature of gardening labor by mine workers, as well as the work of Watut community members to plant on behalf of their host. Some claim that this difference is the result of differential claims on land and efforts to maintain rights in disputed areas. By contrast, Elauru households maintain a closer working relationship as they garden, with more opportunities to cooperate and plant in proximity to other community members.

While gardening remains important, different household experiences shape relationships with the land, the mine, and the labor of gardening. An account of specific households highlights the rhythms of community gardening practices over time. Gardening was again assessed between May and August of each year, just prior to the yam harvest and during the middle of the coffee season. While not comprehensive, these cases exemplify the experiences of families from both villages.

ELAURU

In Elauru, I consider the households of Giamec and Mobi, and Yansom and Sepora. They represent families that don't directly benefit from mining and

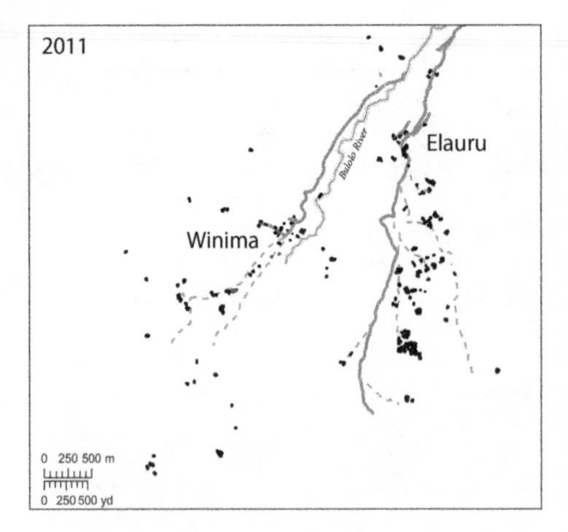

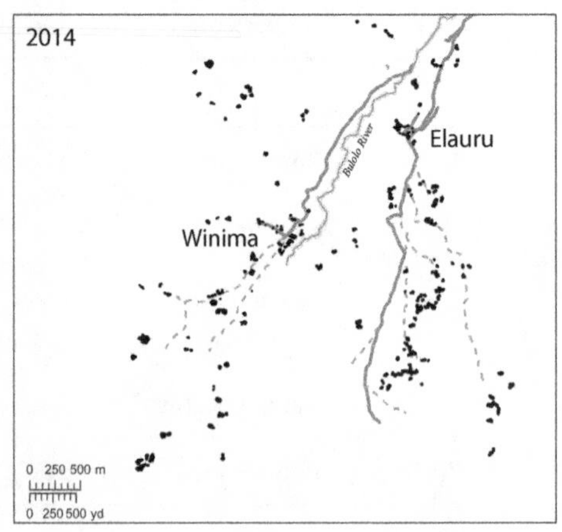

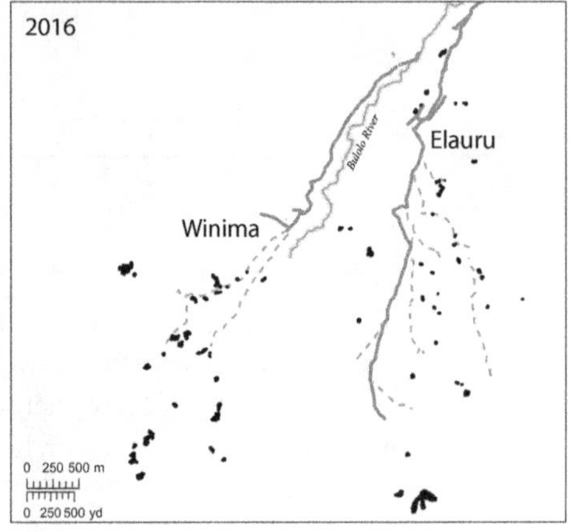

MAP 3.2 Sweet potato and yam gardens, 2011–16. Gardens to the west are associated with Winima; to the east, with Elauru. Note the changes in both size and number of gardens associated with each village. Larger garden plots typically represent multiple households working in close proximity.

highlight how different households practice gardening in a non-landowner community. The first family, that of Giamec and Mobi, is a larger, multigenerational household, with many adult children who can help with garden work. The second couple is older and has passed on many of the responsibilities for the land to the next generation.

Giamec and Mobi, Elauru Village

Giamec and Mobi raised a large family whose residency in Elauru continues to revolve around and support their parents. Their two adult sons and two adult daughters remain closely involved in managing the family land and resources. In spite of their large family, they share a single house. They differentiate primary rights to different areas of land (especially coffee), but the entire family shares a common *haus kuk* (cook house) where they eat their meals. In spite of grade 10 and grade 12 education levels, only the eldest daughter and one of their daughters-in-law work as village-level elementary school teachers (K–1). The pay for this work is minimal and irregular, as the government often fails to distribute salaries.

Because they lack reliable income or access to royalties, gardens are clearly important. Not only do they plant a large area (table 3.2), but they spread their gardens across different ecological zones. As with other areas in Papua New Guinea (Waddell 1972), this ensures against landslides, microclimates, and ecological variability. But they also do so to activate the rights of both men and women in their family. Giamec and Mobi consider themselves to be hard workers, and when faced with the declining viability of gardens in 2015, they intensified their coffee efforts and shifted to becoming laborers in the coffee gardens of others.

During the coffee season the family often worked in Winima picking coffee cherries for a single payment (typically 100–200 kina) or paid by the kilo (typically thirty toea/kilo). This intensified in 2015 and 2016, when sweet potato gardens were dry and unproductive. At the same time, they

TABLE 3.2 Giamec and Mobi's gardens

	2001	2011	2014	2015	2016
Household size	8	11	13	13	13
Yam mounds	102	99	55	122	206
Other gardens (m^2)	1,755.8	2,304.46	2,601.11	1,175.74	1,174.9

emphasized yam production as a response to the drought. Giamec explained to me that his family is "fit to pick coffee," and he hopes other Winima will see this and request their help.

Yansom and Sepora, Elauru Village

Yansom and Sepora are community elders whose children were starting to marry when I first met them in 2000. In 2001 their household consisted of themselves, their adult children, and their children's spouses. Though their eldest son had moved out with his family by 2011, after their daughter passed away they took in her children, as well as the children of another relative after the mother died. During the last half of 2015, when the community was busy planting yams, Yansom died. As he was a central elder in the work of yams, he had planted with many in the community. Work on yams by everyone in the village ended with his death. Though he had harvested eighty-seven yam mounds in the previous months, he planted only fifty-one in the month before his death (table 3.3).

Yansom was known as a hard worker. His wife complained to me often that he would get up, drink tea, and while she was still getting ready for the day, take off for the gardens and return on his own. Sepora often stayed at the house to watch her younger grandchildren and the coffee that was spread out to dry in the sun. Almost every morning she fried flour balls to sell, using the proceeds to replenish her flour supply and buy some store goods. They also paid for any they consumed as a household to avoid ruining the business. Both were proud of the fact that they spent little on store-bought items, preferring to save their meager incomes from coffee. They were so successful at this that they were able to build a house with store-bought materials, slowly completing it over a five-year period between 2000 and 2005. Some days they worked together in their gardens, other times they went their separate ways. In 2014, Yansom talked about how he had prepared a large yam garden by himself. Once the land was dry and burned, he would designate blocks

TABLE 3.3 Yansom and Sepora's gardens

	2001	2011	2014	2015	2016
Household size	6	6	6	5	5
Yam mounds	75	88	87	51	0
Other gardens (m^2)	3,731	1,295.64	1,277.57	1,267.81	398.58

for others to plant. He had planted this garden before with his father, and then once as an adult. I too had planted in this garden in 2001 with Yansom, inserting one of the yam seeds into the mound with magic to help it grow.

The rest of their gardens saw a decrease as their adult children moved out, but there was a great deal of consistency in the area planted until his death in 2015. Planting in the kunai alongside a distant cousin from the same *solonarik* designation, Sepora relied on a smaller area in 2016. In part, she was still in mourning and found working in her gardens too evocative of her recently deceased husband, as his presence continued to inhabit their shared land.

WINIMA

In Winima, Kency and Karibi, and Peter and Iriba typify the kinds of households that remain. Many households have moved entirely to the urban and periurban areas, but those that remain are smaller, often consisting of people who are not in private schools or working for the mine. Kency and Karibi are full-time caretakers of their extended family's lands, while Peter and Iriba only recently moved back to the village. Together, these households highlight different aspects of mobility in the mining economy.

Kency and Karibi, Winima Village

Karibi and her husband, Kency, were among the most consistent presences in the community of Winima during my research. Kency, who had met Karibi through the Church of Christ, was originally from Madang. They settled in Winima, where they were both active in the community and their church. During the exploration phase at Hidden Valley, Kency worked for various companies that did drill tests providing essential data about the ore body at Hidden Valley. It was not regular work, a few months on and then many months off as labor opportunities rotated through the community. During his absences, Karibi's large extended family helped them maintain their coffee and subsistence gardens. Once the mine entered construction, Kency decided to stop working there. The schedule was too demanding and distracted him from his church and family. As the mine went into production, much of their extended family moved closer to town, where they could more easily work at the mine and businesses that supported it. In their absence, Kency and Karibi looked after the coffee of Karibi's brothers and sisters. Members of the Church of Christ and Watut families that her father had sponsored continued to help with occasional labor needs. This also included church youth groups from nearby Kudjuru who were paid to prepare large gardens for several of the Winima families in 2015 and 2016.

TABLE 3.4 Karibi and Kency's gardens

	2001	2011	2014	2015	2016
Household size	6	9	1	2	2
Yam mounds	28	86	111	100	73
Other gardens (m^2)	204.47	364.72	259.02	955.32	1,053.10

Their gardening practices reflect a typical model in Winima, where younger generations migrate to school and work while their parents maintain a presence in the community. Like other households, the number of residents varied greatly over the years (table 3.4). In 2011 they looked after Kency's brother's children for a short time. Things changed dramatically in 2014, as all of their extended family had left, and their children were in school in Lae and Madang. They were also among the last of their family in the community after her siblings all moved to Wau or Lae. During my research in 2015, only Karibi lived at the house in Winima, while Kency spent time with two of their children in Madang. He returned the following year, and they lived alone most of the year.

Benefiting from compensation and royalty payments from the mine, with some shared benefit from investment in Karibi's brother's transportation company, they were fairly well off by community standards. Throughout, they planted what they could, increasing garden area in the last few years as they provided extra food for church events and for their family in Lae. It was quite common for one or both of them to catch a ride to town or all the way to Lae, loaded down with bags of produce to give to their family. They have consistently been among the most active yam gardeners in the community, demonstrating the ongoing importance of the crop.

Peter and Iriba, Winima Village

Peter and Iriba are perhaps more typical of the many Winima households that directed the benefits from mining toward an urban lifestyle. I first met them through my host family in Elauru, to whom Peter is related. In 2000, I and my brothers helped him take apart and transfer his house from the old Winima village to the new one. Between 1997 and 2004, he worked full-time at the mine as part of the security force. Leaving this position, he became a board member for Winima Investment (the landowner company set up with support from the mine). By 2011, Peter and Iriba were part-time residents,

TABLE 3.5 Peter and Iriba's gardens

	2001	2011	2014	2015	2016
Household size	6	1	1	n/a	4
Yam mounds	0	0	0	57	0
Other gardens (m²)	115.26	138.94	21.5	21.5	2,540.76

renting a place in Wau while their children were going to school. By 2014, they had moved to Lae, where they lived in a settlement with other Winima residents. One of their sons or a Watut child they had adopted would stay in the village to look after their house and coffee. During the peak of the coffee season, they would return just long enough to complete the necessary work. To my surprise, they had moved back to the village full-time in 2016. Iriba explained that she was tired of living in town: too many pressures, troubles with drinking, and not enough to do. She preferred working in the garden and tending to her coffee. She explained that now that the kids were older they could return. The mine was also struggling, and like many from Winima, Peter and Iriba faced an uncertain future. During my research in 2016, they lived with two of their older children, who were helping with the coffee harvest.

While they continued to tend to their coffee gardens, other gardens were not used (table 3.5). Throughout, they always maintained a small vegetable garden near their home, relying on the markets in Wau and Lae, as well as rice, bread, tinned meats, noodles, and other store-bought foods. Taro, bananas, and other foods that grow throughout the coffee gardens supplemented their diets on trips back home. Returning in 2016, they planted robust gardens. But moving back also gave them some perspective on how things had changed. Peter complained that the expectation and practice of group work was disappearing from Biangai communities. He thought that Elauru still maintained it, but Winima and Kwembu were "lost causes." For much of the coffee season, he relied on Elauru residents to come and tend to the trees. Some were family members, related through his mother. Others were paid for their labor. In Elauru, his family would complain about these requests for cooperative labor. They too were overwhelmed with work, and while they didn't mind helping, they also knew that Winima families would rarely reciprocate. For Peter, this change in group work extended to other aspects of village life: as he put it, "too many work independent" and don't think of the community.

While trends are difficult to discern, there were some significant changes in how each community practiced gardening. In Winima, paid work groups supplemented some gardening activities, which made the work of planting while on break from the mine easier. Some of the work groups were from neighboring villages, changing the relationships between communities. Furthermore, the use of Watut families to help with major tasks was common. These trends are even more prominent in coffee, as discussed below.

For Elauru, environmental conditions figure prominently in the decline in gardening. During the El Niño event of 2015–16, not only did they shift where they planted, but they also placed greater emphasis on yams. In the case of one family, for example, their sweet potato gardens were half the size of other years, while they doubled their effort at yam gardening. This was a general trend; in a single community garden some distance from Elauru, there were forty-four named plots containing a total of 1,961 yam mounds (see map 3.2, 2016, bottom right). When discussing the 2016 yam gardens, men and women laughed at how much they planted. Frustrated and a bit scared of their dwindling sweet potato gardens, which had been severely damaged during the drought, they saw this distant yam garden as a "true garden," one that would provide food for them when the drought ended the following year. Yams have always provided for Biangai; they were at the very heart of their social relationship with the each other and the land.

With extra income, Winima villagers are clearly able to access other food sources.[7] Elauru villagers joked that the Winima would suffer if rice were not delivered to one of the village stores. My friends in Winima often concurred, noting that "we are a bunch of rice eaters." Rice could easily replace sweet potatoes and yams, funded by mining royalties, incomes, and the various business investments that community members have made since production began at Hidden Valley. In fact, this conclusion has been reported in other accounts of labor's effects on household gardening (e.g., Meillassoux 1975). Here, however, sweet potatoes, like yams, are still planted with a great deal of interest in recent years. Whereas sweet potatoes seem subject to great seasonal variation, the relative importance of yams in Biangai worldviews highlights how ecological relationships are being maintained and, in some cases, reestablished. Yams remain an important source of identity and sociality, and gardening in general remains robust. While Winima's yam gardening practices have been altered by the inclusion of paid laborers to prepare gardens, there is not yet a hidden wound severing their ecological relations (cf. Berry [1970] 1983). Instead, while they might plant fewer gardens of pure sweet potato, they continue to maintain "essential commitments"

(Cepek 2008) to an environmental identity defined through yams. In the end, they remain committed to place.

While mining labor alters much about how Biangai relate with their environment, planting is still a process of folding land and land rights, bringing past production efforts into present claims on resources. Children still learn their rights through participation in the work, through observing their families and knowing the contours of the land. The land bears the marks of these efforts, both physically and aesthetically, appearing as viable gardens for future generations. As such, place and person remain entangled. The appearances of the past attract Biangai back to yam gardens worked before, to refold the soil and reclaim linkages with events that occurred during their youth. If there is change, it is more apparent in Winima. Winima villagers have not forgotten how to plant and grow a garden. But the way they know and work the land is being differentiated between those practices that express an essential aspect of what it means to be Biangai and those that don't. Coffee offers further evidence of this transformation.

COFFEE

Coffee was introduced into Wau in 1928 as part of the experimental station run by the Department of Agriculture. In 1929, the *Canberra Times* reported that the director of agriculture for the colony had ordered seedlings from the Jamaican Blue Mountain estates for planting in Wau (*Canberra Times* 1929, 4). While the initial intent was to create a seed crop for indigenous plantations, in 1931 Carl L. B. Wilde took over the plantation, producing and roasting coffee until his death in 1950. The Wau Coffee Plantation, as it was called, was known for exporting "the best beans in the Territory" (*Pacific Islands Monthly* 1950, 23). While the crop is predominately *Coffea arabica*, many hybrid species have been introduced in recent years, and its importance has increased among the communities.

Coffee was first planted in Elauru and Winima in December 1955. Approximately five thousand trees were initially distributed among the seven Biangai villages, and colonial officers encouraged local nurseries.[8] Elder men remember being uninterested in the work associated with coffee. They complain that the number of trees was small and the return even smaller. That any were planted at all could be attributed to patrol officers pressing the issue. Assistant District Officer Whitehead described the situation in 1962: "Some reasonable coffee plots were seen, these being a little over twenty two and a half thousand trees in the area (25 per person). This is the result of much pressure and assistance from this office and the Department of Agriculture over the last five years. All villages have communal gardens, each person being responsible for one or two rows of trees. Individual gardens are small and generally insignificant."[9]

While coffee was slow to be adapted by Biangai communities, by the time of my fieldwork it was commonly referred to as "our money." One woman argued that her coffee gardens were her bank, while others likened it to gold. Those who had participated in recent training activities sponsored by the mine referred to it as "red gold," following the words of Monpi Sustainable Services, a Goroka-based subsidiary of ECOM Agroindustrial Corporation (Switzerland). Its novelty means that it is not connected to Biangai mythopoetic identities (e.g., West 2012, 121), except that it is a tree crop planted in the ground. It produces because of Biangai relations to their lands. Thus coffee, which is not consumed in the communities, is partially conceptualized in terms of the market, but through its relationship to garden places, coffee has become part of what it means to be Biangai, even as it continues to reterritorialize Biangai naturecultures.

With the exception of cutting the grass and undergrowth, the work of coffee is shared equally among men and women. For more labor-intensive tasks like cutting grass, groups of male family members are typically recruited. Bending low to the ground, men move in rows across the garden, swinging razor-sharp machetes just above the surface of the ground. This is typically done in the month prior to the coffee season. The trees are also trimmed and pruned, and new plantings are tended to at this time. As the coffee ripens, households organize to start picking the cherries. Moving from tree to tree, laborers hand-select each cherry, being careful to choose only those that are ripe. Loaded in ten-kilogram rice bags repurposed to the task, the cherries are carried to a manual pulper, where they are skinned and washed the same day. The beans are then loaded in string netbags to be rinsed in water, removing the sticky residue. Once washed, the coffee is carried back to the community, where the work of drying can begin. Every day, or at least every sunny day, coffee is spread in a single layer on tarps that take over the entire space of the community. A few families are able to dry coffee under their houses, while some in Winima still maintain drying houses with layers of mesh wire racks built with the assistance of the mine. Most of these houses have fallen into disrepair and remain as dilapidated remnants of corporate social responsibility initiatives. Community members claimed that like many other programs, it was not well suited to community life.

The coffee is handled repeatedly throughout the process, as Biangai remove skins that the pulper failed to remove, discard beans that the pulper accidently cut, and turn the beans by hand or with a rake to ensure they dry evenly. If rain looks imminent, a sleepy village comes alive, with community members folding the tarps to cover the beans or quickly loading them into bags. When the clouds pass, the coffee returns to the sun. This is repeated until the beans reach a satisfactory dryness, with community members testing the moisture with their teeth. Finally, dry beans are loaded into fifty-kilogram bags for sale to local buyers who come to the village, or they are transported for direct sale in town. As men and women in each household maintain rights to specific coffee gardens, most households keep the beans separate throughout the process, with husbands, wives, and older children controlling their own sales.

In contrast to gardening practices that emphasize multiple layers of rights based upon use and residence, tree crops remain with the person who plants the seed. As a result, planting coffee is a form of strategic gardening (Walters 1983), as the trees establish a more permanent relationship between the individual and the land. Thus, Winima's part-time residents retain control over their coffee trees even after taking up permanent housing in urban areas. While disputes might result in trees shifting ownership, the rights are largely alienated to a single individual. For absentee landowners, some return for the harvest and occasional work, while others designate representatives to look after their coffee gardens in their absence. Under Biangai sponsorship, eleven Watut households were established in Winima by 2016, representing one-fourth of active households in the community (by contrast, there were no households in Elauru composed entirely of non-Biangai).[10] These families tend to coffee and other property in exchange for a small area to plant subsistence crops and a small share of the proceeds earned from the sale of the beans. Like the plantation economy of Jamaica described by Delle (1998), these outsiders maintain distinctive residence, either within a single compound adjacent to the land of their Biangai sponsors or on the outskirts of the village. Increased employment opportunities for Biangai have led to an increased reliance on these families. As the children of the original Watut families age, they too establish sponsorships with specific Biangai households.

Originally, as in other Biangai villages, coffee trees were planted in both Elauru and Winima in single village gardens, with the whole community taking responsibility for the garden's care. Individual households were given control over single rows, and these were tended with much reluctance. Patrol Officer Sandell reported figures for each village in the 1969 patrol.[11] Following his lead, I repeated these counts in 2001, 2011, 2014, and 2016. However, my focus was on counting trees that were cleaned and in production. These were trees that were being tended, where the grass was cut and the cherries harvested. Trees that were covered in vines or overgrown with overripe coffee cherries blackening on the limb were not counted. As a measure of labor and attitudes toward coffee, my research assistants and I also distinguished mature trees from immature trees to reflect recent efforts at expanding and planting coffee over the previous four to five years. As with yams and other gardens, the goal was to understand the labor commitments made by individual households.

Table 3.6 reflects a total count per village as well as a per-person count based upon active population in the village. Many of the active coffee gardens belong to absentee Biangai, while the labor is performed by those who maintain residence during the coffee season. The counts in 2001 (and 2014 for Elauru) come the closest to a full count of all trees, as fewer trees were left untended in each coffee grove.

The shifting numbers in part reflect the gardener's response to ecological conditions, climate, and markets. The 1997 El Niño event dramatically impacted Papua New Guinea, with drought spreading throughout the country, including the Wau-Bulolo Valley. Both communities remember the fires, the declining productivity of their food gardens, and their need of company and government support. However, heavy rain in South America devastated the coffee crop for the year and created demand on the global market. Coffee incomes increased greatly in the community, doubling what Biangai had received in the previous year, as buyers offered "a good price." A family in Elauru talked about how that really "opened their eyes" to coffee. They started planting new areas, expanding old gardens, and spending more time tending to their trees (figure 3.3). By 2001, this labor was still registered in the young trees that we counted. In fact, Elauru had almost doubled their coffee trees, while Winima increased theirs by 57 percent.

By contrast, the 2015 El Niño event did not result in price increases. While news of the drought brought excitement to the community, recalling previous events, the global prices did not change significantly during 2016. Instead, the timing of the drought and subsequent rains had a different effect. In 2016, coffee trees carried in abundance, as did other fruiting trees. Yams were robust, and sweet potatoes were as big as some yams. One guava tree that borders on my host family's house was so abundant that fruits just rotted on the ground. In Elauru, the number of productive trees continued a downward trend, decreasing to their lowest level in my research. However, the work associated with those trees was evaluated as being greater than any other year. Winima experienced similar labor demands. One man explained to me that he would start picking coffee on one end of his grove, but before he made it halfway the first trees had started to bear fruit again. He had

TABLE 3.6. Comparison of quantities of coffee trees by village

	ELAURU					WINIMA				
	1969	2001	2011	2014	2016	1969	2001	2011	2014	2016
Total productive trees	2,971	75,108	62,801	70,474	48,176	925	63,199	51,453	34,044	43,274
Mature trees	1,425	38,219	57,009	55,600	40,082	770	40,108	44,691	31,825	39,785
Immature trees	1,546	36,389	5,792	14,874	8,094	155	23,001	6,762	2,219	3,489
Trees/person	—	319.61	215.81	260.05	173.29	—	265.54	156.39	160.58	230.18
Mature trees/person	—	162.63	195.91	205.17	144.18	—	168.52	135.83	150.12	211.62
Immature trees/person	—	154.85	19.90	54.89	29.12	—	96.64	20.55	10.47	18.56

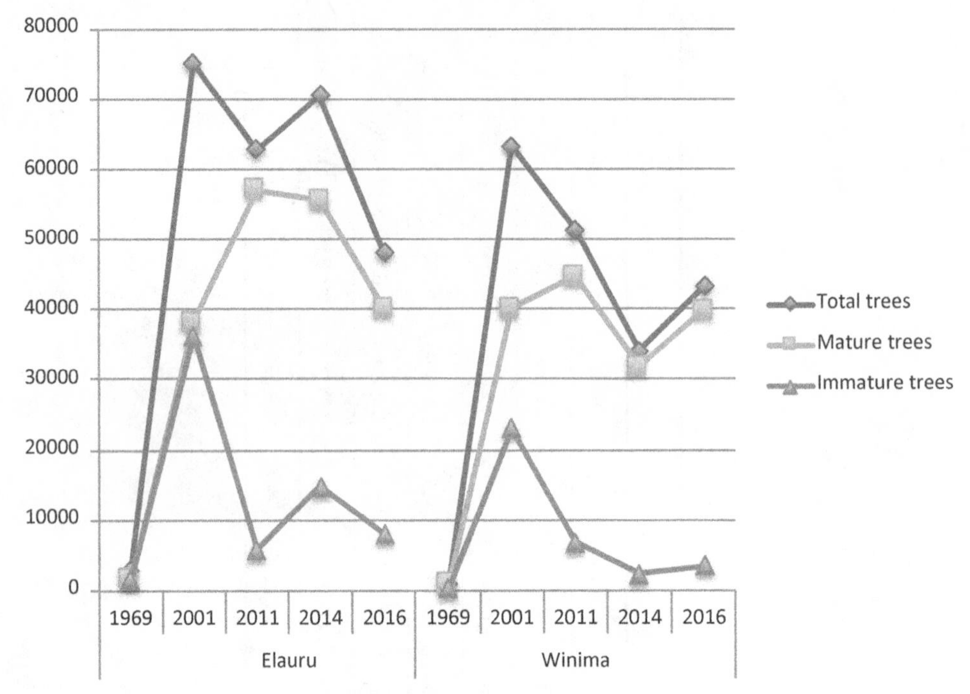

FIGURE 3.3 Comparison of trends in coffee planting between Elauru and Winima

picked three or four rounds already, and the season was not over. This was considered unusual. When I asked another woman about how productive her trees had been this year, she just laughed and said, "We just have to thank God for sending us this drought" because it had made the trees ready.

Between 2011 and 2016 Elauru's mature trees declined in number. In part, this was a shift away from more distant gardens, but more significant was the damage caused by a new drainage system added to the government road. Starting in 2012, the local member of parliament had invested his Electoral Development Funds[12] into fixing and extending the old logging roads that connected more distant communities along the Biaru and Waria Rivers to the coast. But as they refined the drainage to protect the road, water heavy with sediment rushed down the slopes where numerous families had planted their coffee. The sediment settled along the way, and the gardens that lined these waterways were flooded. Coffee was either washed away or suffocated under the excessive water. One older man lost 1,422 trees as a result. The loss was felt deeply, as the first trees in his garden had been planted by his father's hand. In mourning, he left the village for a year to live with a son

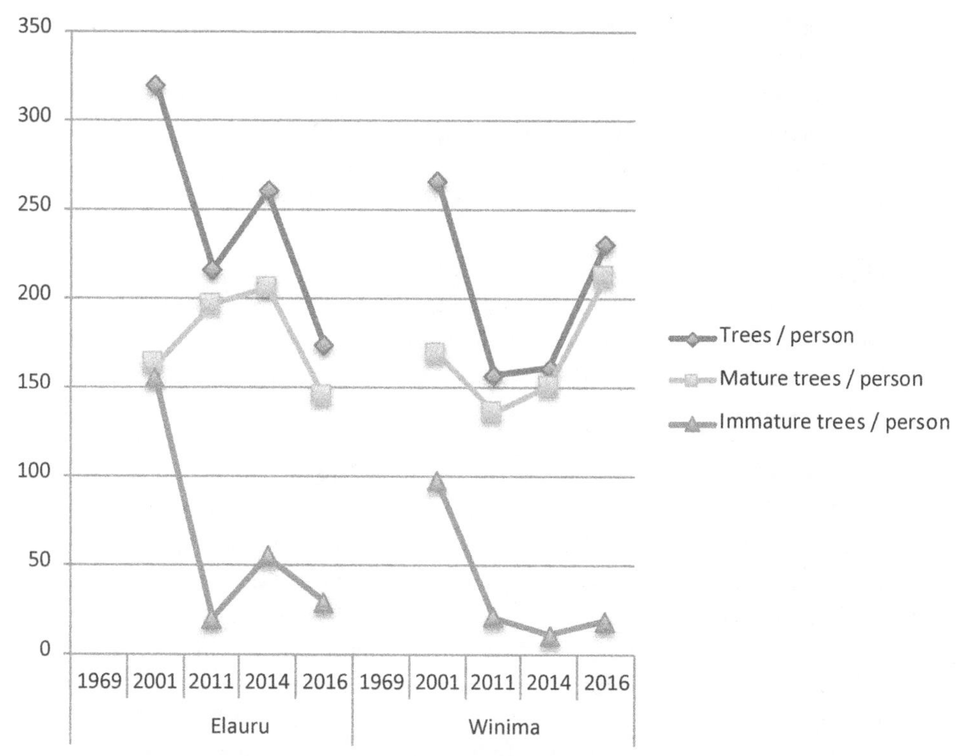

FIGURE 3.4 Comparison of trends in coffee planting relative to the resident labor population

who had married into Wandumi. Like yams, coffee connects gardeners with the land, with broader ecological forces, and with appearances of the past in the form of a parent's labor. But coffee is also money.

Over subsequent years, the early enthusiasm for planting new trees has declined in both communities, and residents are not as invested in harvesting as they were prior to the opening of the mine. Families in both communities commented on this, mostly by publicly shaming those who spend their days playing cards, resting for work at the mine, or spending all their time going and coming to Wau and Lae. But another trend is apparent. Comparing mature trees in Elauru and Winima, while labor effort increased slightly over the past five years in Winima, Elauru seems to be declining (figure 3.4). Crop loss due to environmental conditions is a major factor, but there has not been a subsequent effort to replace lost trees. Instead, many of the most distant gardens have been abandoned, with only a few remaining committed to working along the forest edge. Winima, by contrast, has increased the number of trees in production as the resident population has declined. However, a major factor in this transformation is the increased reliance on paid labor. In addition to Watut laborers who reside in the village, Elauru residents are decreasing their labor effort on their own gardens while increasing their presence in Winima. This was most apparent in 2016, as on many days two or three Winima families were hosting Biangai from Elauru who were working Winima's groves of coffee. Coffee was not just money; it also fostered relationships that were market based, as labor was exchanged for income. During a church meeting in Elauru, community members set a rate of fifty kina to help families in Elauru, and one hundred kina for families in Winima. They felt this was a fair price, as Winima had access to money from the mine.

Like yams and sweet potatoes, coffee is experienced in a very Biangai way, but at the same time, its status as a commodity is not insignificant. Like the Gimi, Biangai see coffee as placing them "within a wider system of relationships and exchanges" (West 2012, 128). It links them with regional markets and establishes relationships with national buyers and global markets, which remain somewhat mysterious. While I would agree with Paige West's overall assessment that coffee in Papua New Guinea communities refocuses "personhood away from collectivity and transaction and toward being individualized, self-possessed persons" (128), Biangai persons only differ by degree and not in kind. Individuality might be seen as emergent or contingent, as practices of managing the land and broader social relations still hold sway.

BECOMING FARMERS

"We need to become *farmers*," an Elauru man repeated several times, emphasizing his use of the English word. We were discussing a series of courses sponsored by the mining company on coffee, animal husbandry, marketing, and business. Unlike other mining benefits, members of all Biangai communities are invited to these training sessions. I noted that I had never heard the term "farmer" used by Biangai. He explained that when he went to a recent coffee course, they said, "You are a farmer." I asked for clarification: "What does it mean to be a farmer?" He responded, "Before we worked in our gardens to grow our bellies; now we want to see money." Farming, he explained, involved not just subsistence crops but a diversity of cash crops and animal husbandry practices. That was why he tried new seeds and new species of plants and animals acquired from government agencies, the mining company, and other farmers. He pulled a plastic bag from his *bilum* and emptied the contents into his hands. "This is a soya bean," he explained, "a money bean," and he devoted one garden to trying it out. For similar reasons he also acquired goats, ducks, and fishponds for himself and others in the village. Now he had to manage all of these different "farming" activities. As a result, he did not plant yams that year because he had "to concentrate on all of these projects."

Courses are one way neoliberal regimes shape stakeholder communities (Welker 2014), refining what it means to be a responsible, self-motivated worker and citizen. As one trainer noted during a business skills seminar, "The training is like a seed and if planted and looked after well will bear fruits one day. The ball is in your court to start implementing" (*Morobe Miner* 2012). Morobe Mining Joint Venture initiated a number of these programs in support of long-term sustainability of communities in the impact area of the mine. For example, Mainland Holdings, the third-largest exporter of coffee in Papua New Guinea, was contracted by the mining company to offer training and serve as an outlet for coffee sales. The company had specific export standards that it expected growers to live up to, and the training was designed to ensure that quality standards were met. Having gone through training with Mainland's specialists, Biangai were expected to meet these export standards and not just the needs of local bellies. These courses offered more than information about coffee production. Specifically, the training program was described in a Coffee Industry Corporation publication as "aimed at enhancing the income of the Biangai community through improved coffee management practices. It focused on improving the skills and knowledge of farmers on coffee nursery establishment and field planting, coffee garden management, basic garden rehabilitation and pruning practices, coffee quality improvement through improved harvesting and processing techniques and basic financial management and cash handling practices and Marketing" (Coffee Industry Newsletter 2011).

Visits to the community by Mainland employees in 2016 emphasized these practices, as they checked on the quality of the groves and gave speeches to the community about how their coffee now had a name in the world market. "You can now purchase Wau Co-op Coffee in Australia, the United States, and Europe," they announced to great applause. As my interlocutor's striving to become a farmer suggests, these courses were initially effective in transforming how Biangai in all the communities viewed their relationship to crops and thus to the broader environment. The techniques for improved quality were driven by industry standards and decades of academic research. The emphasis on financial management, marketing, and crop quality shifted the focus from gardening for subsistence to farming for money. But more than simply reflecting a shift from subsistence to money, they taught a shift from an intimate relational ecology to an instrumental market-based ecology.

Of course the company viewed these interventions as increasing the income of area residents and fulfilling their contractual obligations with the government to create postmining sustainable incomes. In their reporting, Morobe Mining Joint Venture (MMJV) emphasized the program's cost (28,699 kina for "basic husbandry training") and the goal to "increase income levels in the community and reduce dependency on mining activities" (Harmony 2011). In its newsletter, MMJV describes (but does not quote) a female participant as learning "among other skills, proper pruning techniques and picking the right cherries to produce quality beans" (Morobe Miner 2011). Biangai who did not participate continued to produce coffee as before, pruning and selectively picking cherries. They did not see the advantage of the courses; after all, they had been tending coffee since the 1950s. But participants like my Elauru friend felt transformed, enlightened, "like farmers." In 2011, the way he related to his gardens had changed. Another man told me that he had given up yams for the year because cultivating yams was too demanding and took him away from his farm. He imagined when I returned that he would have become a "countryside farmer," using a phrase from a course to demarcate the transformed identity and distinctions in spatial relations to urban areas. When I returned in 2014, he had returned to planting yams but continued to see himself as a farmer. He complained that Biangai had participated in many courses, but few were putting them into practice. He still insisted that he wanted to become a farmer, but what that meant seemed to be shifting away from how he saw himself. Too many distractions and other commitments, he complained, kept the would-be farmer unfocused. He needed to plant yams to fulfill funerary obligations, and markets proved lacking for the many novel crops he had tried. While yams continue to invite some Biangai to dance, it remains to be seen what impact becoming a farmer will have. But even more complicated is the impact that becoming a worker is having on Biangai agriculture and, by extension, Biangai relationships to place and ecological identities.

WORKING AND FARMING

In contrast to economic literature that emphasizes dual economies with marked ideological boundaries between market and subsistence (e.g., Gudeman 1986), I have argued that these practices must be seen in light of broader ecological relations. Biangai practices of working the land, regardless of the crop, require an intimate connection with contemporary and historic social relations. James Leach (2003) distinguishes between formal production (an instrumental relationship to space) and gardening (the process of growing food as unfinished and always being made). There is no end product, only further gardening. While very much akin to Wendell Berry's distinction between having a relationship to place and managing it, Leach sees these as contrary images. He states: "Garden knowledge not only comes out of a history of relatedness and movement, it is the ongoing production of these relationships that a gardener seeks to energise. Thus procedures within the garden mirror the form of relations outside it" (Leach 2003, 102). In the practices associated with gardening, relationships are created and maintained. In Leach's words, they are grown. For the Biangai, a similar relationship has existed between *solorik* groups and their garden lands and between persons and places. Thus, the work of gardening is also the work of creating and maintaining placeperson relationships in place. They are not produced and thus finished, but require ongoing attention and care.

Nonetheless, the multispecies encounters that form the contours of such practices are being reimagined in this neoliberal moment. Mining and mine-sponsored activities alter the physical landscape and reshape the experiences of laborers. They transform ecological relations via corporate-sponsored training and development programs, leading to the *partial* creative destruction of garden life. Decreased production of subsistence crops alongside increased diversification of "farming" activities and opportunities for mining employment challenge Biangai ways of becoming with their gardens, as well as challenging the garden's ways of becoming with Biangai. For sweet potato gardens and groves of coffee trees, there are greater signs of transformation. And yet, Biangai desiring to become countryside farmers do not lose touch with yams. As the evidence so far suggests, yams haven't lost their ability to invite gardeners to dance. Yams remain a robust part of community life and labor, marking and refolding relations to land. They remain essential to a local identity, informing ideas about production and growth as well as broader social and ecological relations. While there are some changes in the organization of labor, especially as it relates to coffee, the practices of yam gardening remain sources of hope and agentive power, connecting Biangai to the ultimate source of their sociality in the land. They are counter to wider ideals of neoliberal subjects, and for now, both communities have found ways to make yam gardens ongoing possibilities. Such values are, however, at stake in the organization of Biangai places for conservation efforts.

CHAPTER FOUR

Becoming Conservationists

FOR more than fifty-five years, biologists have used the Upper Bulolo Valley as a research site, establishing the Wau Ecology Institute on land that once belonged to Biangai speakers. With shifts in the political ecology of this young nation, scientific and Biangai interests have increasingly merged under a rubric of sustainable development but diverged over what is sustainable and what is development. The environment and humans are immersed in different systems of value that emerge through the discourses and practices of ecotourism, biology, and local ideas of place and the environment. In the Upper Bulolo Valley, these systems collided in the initial success and eventual failure of a small-scale wildlife management area. The case study highlights how values are placed and displaced by conservation and biological research practices.

Established in 1961 by J. L. Gressitt, an American entomologist, the Wau Ecology Institute (WEI) worked with communities throughout Papua New Guinea for many years on sustainable development and conservation efforts, while supporting hundreds of biology researchers as they completed their dissertation and postgraduate work.[1] Its midmontane forest location on the island, deemed central to the speciation of ancestral insects now scattered throughout the South Pacific, made it appropriate for Gressitt's interests in Pacific species diversification and adaptive radiation (Gressitt 1963). It served as a base from which lowland and highland species could be collected and observed in their habitats. Though initially established as part of this wider project, the Wau Ecology Institute became an important feature of Papua New Guinea's emerging civil society during the late colonial period. Like other conservation efforts in Papua New Guinea,[2] conservation was presented as a sustainable form of development, one that could forestall the extractive practices of capital while providing access to the benefits of modernity.

While scientific entomology and other biologies figured strongly in its program, WEI's projects diversified into an increasingly applied focus on sustainable agriculture, environmental activism, and local empowerment. Nationalized at the time of Papua New Guinea's independence in 1975, WEI developed a set of sustainable development projects to establish its own financial independence: organic shade-grown coffee was grown and milled on WEI's property, a guesthouse offered a place to sleep for visiting scientists and tourists, and grants from Japanese, American, European, and Australian NGOs and governments funded a variety of other activities. They also continued to support research into medicinal plants (Wooley and Göltenboth 1991), sustainable agriculture (Göltenboth 1990), and unsustainable logging and mining (Sakulas and Tjame 1991). A project was developed to purchase and market insects on behalf of a wide range of communities throughout Papua New Guinea, though largely focused in Wau. Harry Sakulas, a Papua New Guinean academic and former director of WEI, has described the activities of the institute as emphasizing the integration of programs and practices: "Research is closely linked with training, community extension, management of economic ventures, campaigning and policy making so that all projects are interlinked as part of an integrated program" (Sakulas et al. 2000, 93). Grants also funded a number of conservation efforts throughout the region, including work in the Kuper Range on land belonging to Elauru village.

Two aspects of WEI's conservation and biological research program are of interest here. First, values are at stake when global ecological ideologies intersect with local values of place, persons, and wildlife. While often touted as an ecofriendly business (Small 2007), dead stock trade in insects emphasizes the movement of insect bodies from places threatened by plantation economies and extractive industries to private collections and museums, compensating local collectors for maintaining insect habitats. In contrast to other animal species, whose economic worth for tourism is higher in situ, insects are highly collectible. The insect market is celebrated as significant for both sustainable development and habitat protection, achieving the somewhat stable alignment of a species's economic value and intrinsic value for entomologists. However, these microfauna transform spatiotemporal relationships as their bodies are "mined" for the international insect trade. Second, a relationship is evolving among the community of Elauru, ecotourists, and scientific researchers, as Biangai host them in their own conservation facilities, work with them as guides, and debate the values of conservation as a place-based practice. Both cases highlight the ecological and material networks that are supposed to make conservation and biological research possible, and the collision of persons, places, and values that challenges the long-term success of such strategies.

DEAD STOCK AND CONSERVATION

In 2001, Gilles Deslisle, a Canadian art teacher and amateur entomologist, was convicted of illegally importing six specimens of *Ornithoptera alexandrae*, the Queen Alexandra butterfly, from Papua New Guinea. With a wingspan of twenty-eight centimeters, it is the largest butterfly in the world. When the naturalist A. S. Meek first encountered this butterfly, he mistook it for a bird and shot one for the British Museum, where the specimen remains today (Freeborn 2014). In 1992, with the help of AusAID, the Australian government's aid agency, a conservation area was established in Oro Province (some distance from the Bulolo Valley) to protect the species (see Barker 2016). As with other conservation areas, called Wildlife Management Areas in Papua New Guinea, the project sought to encourage ecotourists and scientific researchers to use the area, finding alternative values in the forests in order to forestall the development of logging and plantation economies.[3] While a number of insects have been marketed as part of this project, the Queen Alexandra serves as an umbrella species for the area. However, it is illegal to export the species, dead or alive. Listed in appendix 1 of the Convention on International Trade in Endangered Species of Wild Fauna and Flora (CITES), it is considered one of the most endangered butterflies in the world. According to current regulations, it is to remain in place.

Mr. Deslisle did not act alone. He worked with a local man, Russell Hauro, who prepared the specimens for shipment to Canada. Russell runs a guesthouse where Gilles stayed in 1997. During his stay they made arrangements to collaborate in the illegal shipment of a variety of insects between 1997 and 1999. The two were caught by Canadian customs in July 2000 after a multiyear investigation. Deslisle was charged and convicted, while Hauro was granted immunity in Papua New Guinea so that he could testify in Canada. The dead value-laden bodies of the insects were to be destroyed. In Papua New Guinea's newspapers the motives for the scheme are apparent. Deslisle is described as "an illegal trader . . . wanting to exploit PNG's rich flora and fauna" (*Post-Courier* 2000). For his part, Mr. Hauro "was paid between US$200 and US$300 for a butterfly, which fetched an estimated US$10,000 and US$20,000 when sold on the black market" (ibid.). As reported in the Papua New Guinean press, the values placed on these objects in motion were monetary.

But the case is much more complicated, and more than money displaced these insect specimens. Court records detail a different set of values—values that not only offer an alternative to the public narrative of greed but also serve to separate and mark persons and places connected to the Queen Alexandra species. Deslisle had published a number of articles on regional insects, including a proposed reclassification of the species in question (Deslisle 2004). On this basis, he made his case in court for a scientific interest in the specimens, which he argued absolved him of wrongdoing. The values of science, he seemed to suggest, are pure and subsequently purify his activities. He had no intent of reselling and noted that the six specimens in question hardly impacted the population. The court found that his intentions seemed genuine, and he was able to convince the trial judge that money was not the motive here. However, he had still broken the law. Before upholding the lower court decision, an appellant judge summarized the case: "In her reasons for the sentence, the trial judge noted that Mr. Deslisle had made the study of butterflies his life's passion and that he had authored several articles on the subject. She was satisfied that he had not imported the butterflies for commercial purposes, but, rather, to further his reputation as a world expert on them. In essence, she found that his motive was *ego satisfaction, rather than monetary gain.*"[4]

Furthermore, court records document a much larger transfer to Hauro, in the nature of C$30,000 over the course of two years to support the conservation effort as the international funding was ending. In spite of Hauro's apparent financial success, the trial judge viewed him as a simple man: "An exceedingly unsophisticated individual who, in effect, was transported from the bush in Papua New Guinea to give evidence in my court" and who "succumbed to the temptation of the large sums of money." Thus, a different set of positions emerged: the career-driven Deslisle motivated by the quest for knowledge and the desire to enhance his reputation, and the unsophisticated native taking advantage of the offer.

Is this the end of the story? Does it matter that Hauro had been an active proponent of the Oro Wildlife Management Areas, involved early on in identifying and planting vines favored by the species in question (Hibberd 1997)? Or that the AusAID funding was ending, and that funding paid a great deal more to expatriate workers than to their local counterparts (Aidwatch 2005)? Or that Hauro and others were to be left without support as the project lost financing?[5] Did the judge consider how Hauro might value the species and conservation, not to mention how he valued the places where he lived and cared for his family? Furthermore, it seems that both were well versed in debates over whether opening up the market to sell Queen Alexandra specimens would do more to save the species than to hurt it (e.g., Parsons 1992). Could both individuals be motivated as much by species protection, science, intrinsic valuations of place, and the social relations they established during the course of their exchange? Or is it as simple as the judge concludes? These questions are left unanswered in the public account.

Of interest here are the different ways in which the movements of a single species of insect inform distinctions between varied human-environment relations, as well as the way insect movements carry values between places. The Queen Alexandra is differently marked as an object of scientific study: an endangered species protected and restricted at the highest levels of national and international authority; an object of individual self-actualization, of biophilia (perhaps) and most certainly passion; a symbol of a province, a region, a conservation area, and a community; a part of a regional resource management strategy; and a significant aspect of an indigenously understood landscape. Differences are highlighted between legal and illegal, sophisticated and unsophisticated, the Papua New Guinea bush and urban Canada, courts and conservation areas, scientists and collectors, and so on. Throughout, the Queen Alexandra is intimately tied to and makes the place that came to be called the Oro Wildlife Management Area, even as it travels.

Ultimately, it is the entanglement of species in human-environment networks and the play of values as species shift from place to place that makes their movements meaningful and possible (Tsing 2015). Of particular interest is the work that flora and fauna are expected to perform in the development of conservation areas. But in contrast to the rhetorical emphasis placed on maintaining and protecting a species in situ, much of the conservation efforts in the developing world require that specimens or parts of specimens be in motion in order achieve conservationists' goals. For example, collectible insects that move as part of the international dead stock trade are not seen as a threat to biosecurity, monocrop agriculture, or national borders. Instead, the conservation and monetary values of insects in motion increase as "specimens" are killed, prepared, and sold. In the process they are envisioned as raising funds for and awareness about conservation. Likewise, scientific knowledge, photographs, and samples compose the traffic of values for many other projects. Wau Ecology Institute just happens to include each of these in making the Kuper Range legible.

In Papua New Guinea, insects were sold through a government agency (the Department of Conservation's Insect Farming and Trading Agency) and designated NGOs (such as the Oro Wildlife Management Area or the Wau

Ecology Institute). These organizations provided an outlet for the movement of scientific collections, gathered during government-approved entomological fieldwork, which are locally collected and funneled into private collections, labs, and museums. The Insect Farming and Trading Agency, though no longer active, sought to achieve Papua New Guinea's unique constitutional goal of protecting insect species (perhaps the only nation to include such provisions) by facilitating the movement of the dead-object bodies of insects. But insect movement for conservation also includes the value attributed to them during scientific research, where they are photographed, distributed to multiple museums and collectors, selected as type specimens for collections, dissected in the study and differentiation of insect class and kind, and made and refashioned in light of local values and other entanglements. While such movements might make Western-style conservation *seem* possible, they negatively transform the moral landscape of human-environment relations. In other words, through their dis*place*ment, conservation insects maintain more than the borders of Wildlife Management Areas (Halvaksz 2013a); they introduce and reinforce a set of human-animal/human-environment relations that are conducive to extractive thinking and the disentangling of place and person.

INSECT VALUES

Among the many protected species in Papua New Guinea, insects make differences salient even as they enlist human agents in the common causes of their collection and eradication. They have the capacity to charm us through their startling colors, to kill us with a bite, captivate through their organized labor and their collective works, and more often than not, bug us and our theories of human-environment relations. As bodies that move under the power of their own volition and with the help of human hosts, insects also provoke, pollinate, wriggle, and tickle other values.

Value is not to be taken as simple economic value (see Losey and Vaughan 2006). As David Harvey (1996, 155) notes, monetary value has the power to liberate, but it fails to capture the complexity of human-environment relations because monetary valuations limit how we understand the environment and relate it to an instrumental value. Nor are insects simply placed within a utilitarian framework, where their intrinsic value as part of living spaces is emphasized as central to maintaining "ecological integrity and health" (Samways 2005, 9). Insects lend themselves to these meanings and more as they circulate from place to place.

In a critical review of nature's values, geographer David Harvey distinguishes the emptiness of capitalist values from other ways in which nature comes to be known, noting that all values—scientific, religious, personal, metaphysical—"rely heavily upon particular human capacities and particular anthropocentric *mediations* (some even upon the charismatic interventions of visionary individuals)" (1996, 158; see also Halvaksz and Young-Leslie 2008). Harvey's argument is that such diverse values are suppressed and supplanted by monetary values under the press of capitalism.[6] While gold mining certainly presses upon local place-based values, species values are not so simply distinguished (see Mitchell 2002). As the Queen Alexandra case demonstrates, while the six specimens came to represent money in the courts and media, they remained tropes of conservation areas imbued with human passions and scientific meanings. Their movements were as much about the production of place as they were about the celebration of specimens.

In Oceania, human-animal relations are seen as mediated through place (Roberts et al. 2004), creating webs in which conservation specimens, places, persons, and values are entangled and engaged. Philosopher and anthropologist Bruno Latour notes that the objects of scientific inquiry and knowledge "are not found at the *meeting point* between things and the forms of the human mind; phenomena are what *circulates* all along the reversible chain of transformations, at each step losing some properties to gain others that render them compatible with already-established centers of calculation" (1999, 71–72; emphasis original). Like gold and other minerals, values attributed to conservation specimens are gained and lost by different stakeholders as the insect bodies move, but emplaced values are always revealed in the trace of their trajectories.[7] In a context where local populations directly intersect with scientific research, we have to consider the dialogic production of knowledge and values—that there is a reverse trajectory for the generalizations that renders phenomena compatible with indigenous "centers of calculation." How conservation specimens are set in motion matters as much as how they are collected and consumed in scientific and private collections, because ultimately species thrive in their habitats and not in labs and collections.

As part of the swarm, insects embedded in large moving colonies carpet the surfaces of conservation and burrow deep into its grounding. Supposedly, their numbers make them impossible to kill off without destroying the habitat itself (Orsak 1993). They are the habitat—consuming, decomposing, and tending to places and placepersons who dwell among them. Removing the flora and fauna with which they live effectively removes an insect population, but removing one specimen, or perhaps hundreds, does little to destroy the group. Instead, it connects the very partible swarm to an extensive network of values and places. In contrast to the multimillion-dollar insect farming industries, where insect bodies are produced in mass for decorative trades, conservation insects are conceptualized as inhabiting free-range ranches (Small 2007). The movements of the six specimens discussed above thus inform how values are constructed within and about a place in Oro Province, as well as courtrooms in Canada. Likewise, the movements discussed below speak volumes for conservation practices, practitioners, and places associated with the Kuper Range, and help us to understand why the effort to designate an official Wildlife Management Area ultimately failed.

HUMAN-INSECT RELATIONS ALONG THE UPPER BULOLO

While insect narratives of both communities inform my work, the present discussion develops from Elauru's efforts to establish and maintain a conservation area, and the subsequent destruction of the research station buildings in 2005. Like the Wola, Biangai "evince relatively little interest in insects on the whole" (Sillitoe 2003, 85).[8] Biangai do not name every species, with many referred to under a more general category, *wetagorugoru wiri*. But even the nonspecified species figure into their daily lives. For example, they are central to Biangai horticultural practices as both pollinators and invaders. Gardeners preparing a new expanse of secondary forest for planting will burn the dried remains of cut flora to return nutrients to the soil and kill off beetles, grubs, ants, and what Biangai generally refer to as *ol liklik snek* (small snakes), which live in the ground. Secondary forests are preferred over kunai grasslands, as they provide the greatest "heat" for this eradication. Fires are also lit at the base of growing yam vines to heat the ground and smoke the leaves. Both are done in part with insects in mind, affecting a local model of biosecurity grounded in the borders of subsistence production. As yams grow in size, so does the noted skill of the gardener. Insects that succeed in consuming these prestigious tubers effectively consume the gardener's stature.

But not all insects are associated with crops. Biangai have historically eaten certain insects, and they interact daily with a variety of species at a certain level. The skins of *kundu* drums are kept in tune with the wax of the *meyak*, a type of stingless bee (probably *Trigona* sp.). And the nightly cry of the cicada (perhaps *Pomponia* sp.) at dusk remains a pervasive marker of Biangai temporality. Biangai insect-human relations are to an extent gendered and generationally distinct. Children of both sexes are more likely than adults to consume grubs and beetles during their travels between village and garden spaces, and they are even more likely to incorporate insects into play. Beetles and butterflies become the equivalent of remote control airplanes, tied to their human pilots by thin twine until their deaths. Women too consume wood grubs when found in abundance, particularly during the later stages of pregnancy, when such foods are believed to aid labor.

However, none of these invertebrates figure into how Biangai conceptualize their own identity, mythology, genealogy, and personal histories. None of them are collected. Biangai certainly have knowledge of insects, but they are distinct from other animals, for which Biangai show much greater affinity. The tree kangaroo, cassowary, wild dogs, pigs, eels, and many species of birds and snakes are important in Biangai exchange and subsistence. Furthermore, these animals are central to Biangai identity as features of myth, representing genealogical connections, paths (*kasi mek*) and affiliations. Insects, by contrast, are often described as *samting nating* (something nothing). This is the context for entomological narratives of insect values practiced by WEI.

CONSERVATION INSECTS

Planning for the Kuper Range Wildlife Management Area began in 1989 between WEI's director and Elauru villagers on land controlled by the community. In particular, then director Harry Sakulas began discussions with various community representatives regarding WEI's long-running Kaindi research station, which was located adjacent to mining leases and squatter settlements on the ridges above Wau. He hoped to move the research station twenty kilometers farther up the Bulolo River, away from the pressures of mining. Elauru villagers expressed interest in setting aside some of their lands for conservation to replace this site. The Conservation Needs Assessment for Papua New Guinea (Beehler 1993; Sekhran and Miller 1995) identified forests in the Kuper Range as a biodiversity hotspot, and the region of the proposed camp was also a significant site, once inhabited by malevolent spirits that Elauru religious leaders had prayed away in previous years. In 1992, WEI constructed a research station on the five hundred hectares of land offered by one *solorik*, moving its increasingly overrun Kaindi Research Station buildings to the site. The immediate use of this facility by tourists, researchers, students from the Bulolo Forestry College, and a *National Geographic* film crew increased village interest in the project, and other *solorik* groups contributed approximately one thousand hectares of additional land in 1995. In retrospect, Biangai consider this period in the development of the conservation area as its heyday. A number of grants funded biological surveys and mapped the boundaries of the area (Halvaksz 2013a). Income was generated, relations were established within the region and with international visitors, and Elauru residents seemingly enjoyed the rhetorical strategy of celebrating conservation over mining and logging. As Sakulas and his colleagues have argued, "Alternative income sources are demonstrated to the villagers by action rather than theory" (2000, 95). Insects were envisioned as one such alternative.

Wau Ecology Institute's investment in the insect trade developed in tandem with its increased emphasis on sustainable development and conservation. Under its integrated approach, staff of the Butterfly Ranch (as the division was called) provided local training and equipment, marketed the specimens overseas, and arranged for the CITES export clearances. The project was initially envisioned as providing an alternative income for local communities, as well as funding operating costs for WEI. Director Harry Sakulas (1998) emphasized these monetary values, noting that the international trade in live and dead stock for scientific and private collections as well as decorative industries generates billions of dollars a year. For WEI, the goals of the insect trade combine "economic incentives" for community participation and scientific values "as villagers have been encouraged to bring anything unusual and interesting. In this way this has in turn given an incentive to the village elders to pass on knowledge of particular species, their feeding plants and breeding cycles to their next of kin" (Sakulas et al. 2000, 95). In Sakulas's idealized vision, local collectors embody diverse values ranging from local knowledge transmission between generations to a globalized perspective on the significance of a species.

In 2001, the manager of the Butterfly Ranch, Michael Hudson, described to me the trade as practiced at WEI. When requested by a community or group, his staff provides training in the necessary methods, including how to identify collectible species, define their habitats, increase insect food sources, and finally collect, kill, and package the insect bodies for shipment to WEI, where they are marketed to overseas buyers. Once sold, the collectors are paid for their efforts. Compensation is paid only for those that are marketable. Damaged, faded, imperfect, or overly abundant specimens are discarded. Village collectors quickly learn the market sensibilities of overseas enthusiasts. Though instruction is given, the qualities sought by collectors are learned through trial and error, transferred with the insects' movement. Buyers repeat good exchanges, while bad experiences mean that insect enthusiasts will not order again and may report their troubles to online bulletin boards (see insectnet.org, for example). Community participants quickly discover that some specimens are worth the effort, such as jewel beetles (family *Buprestidae*), various butterfly larvae, and other colorful specimens. Insect values are transformed in the process, increasing the demand to plant insect foods, encouraging more careful specimen selection, and revaluing what might once have been simply "something of the forest" into something worth money.

In-village training focused on developing values: "To further strengthen and encourage the industry, training concerned with quality control for the international market is provided. Field officers who return from their buying trips recommend that keen farmers attend work attachment and training courses on collecting, drying and packaging, at the Institute. When the farmers return home they not only concentrate on looking for *highly priced* insects but know how to deliver *good quality products*" (Sakulas et al. 2000, 96; emphasis added). While much of the insect conservation discourse emphasizes the gains in terms of the environment, poverty, science, and aesthetics, what is clearly being performed in this conservation habitat is the creation of a different set of values. Insects are transformed into *good-quality products* that can earn a high price. Ideally, this in turn transforms the value of the habitat into *good-quality valued places.*

Elauru villagers were trained in these practices as part of their involvement with WEI in the development of a wildlife management area. It was one of the alternative income sources that would supplant logging and mining. However, Hudson complained that the Biangai attitude in this effort was lackluster. During interviews, Biangai regularly spoke of this work in less than enthusiastic terms, and such practices must be placed in the context of other work performed in the community. Work is a value-laden activity, and willingness to perform certain tasks reveals something about the set of values associated with that task. Working with yams, for example, connects gardeners to places, to ancestry and histories of social relations. Does working with insects make the conservation of places locally legible?

As part of my larger research project in 2001, I asked a random sample of community members to rank and discuss thirty-four work tasks identified during previous interviews. Listing and ranking provoked much discussion, as participants rated work tasks according to difficulty. Those tasks that were considered the most difficult included those closely associated with Biangai identity, especially Biangai masculinity. Planting yams, cutting a new area in the forest for a garden, and house construction were consistently rated as the most difficult by men and women alike, while daily household tasks such as fetching water, cooking, cleaning, and doing laundry were rated as among the easiest. The task of collecting and selling insects was ranked twenty-third out of thirty-four but had one of the highest variances, with a range from easiest task (ranked 1/34) to one of the hardest (ranked 32/34). Reasons for this variation reflect degrees of experience with the practices and an outlier who held a distinct perspective on what constitutes hard work.

One landowner in the conservation area was familiar with the practices of collecting and knew all the staff at WEI quite well, yet he had never participated in the insect trade. His description is informative of how this work was locally received: "Okay, selling insects, that is . . . that's one kind of work. It's work because I have to find them, carry them to town, and sell them. That is a small task. It's not like I have to go and search and search and search and fill a large bucket and carry *that* to town. It's like, I find one and put it inside my pocket or inside a plastic bag and take it in." He begins by trying to justify how he might see this as work, breaking it up into the tasks involved. But he concludes by belittling the work: rather than filling a large bucket, it is just slipping something into his pocket. In contrast to other work, this was not really that difficult. The one informant who ranked it as a difficult task explained that he did not know how to do it, while more strenuous tasks, such as planting yams, he considered easy because he had mastered the practice. His rankings are a reverse image of the general trend but place insect collecting in the same devalued position as other respondents described. Insects for him and others were a child's task, "something that children should do to make some money for gum and sodas." Another landowner in the conservation area explained that he thought it was important work, but it was neither too difficult nor too rewarding. "It depends on the money, if the money was enough. . . ." He trailed off, suggesting that he would be more interested if more money were involved. A third man also directly related the ease of the task to the monetary returns. Having attempted to participate in the insect trade for some time, he explained his efforts with some frustration: "Yeah, I used to get, say, six kina, but only after six weeks. It depended on how many you sold, five kina, six kina, or ten kina . . . that's all. But you have to give them to WEI and then come back in six weeks to see if you got any money for the effort." For him, there was simply too much time between the tasks and the payment to merit any long-term attempt.

The effort to encourage Biangai to sell insects faced two essential problems. First, insect collecting was central neither to Biangai identity nor to values associated with work. Adults associate the collection of insects with children, as a form of play. It was easy and therefore not considered "real work." Second, the project was explicitly about connecting monetary values to the conservation area. Insects were to be sold for money, but this value was not easily performed. The discerning overseas clientele of the insect market meant that WEI could not purchase insects outright, delaying monetary returns until the specimens were sold. Amounts paid were also deemed low and uncertain, as one had to wait at least six weeks. Insect movements were not easily monetized, and the value to conservation places was disconnected in time and space. As the values associated with insect specimens in motion toward this end of the chain were generalized within indigenous "centers of calculation," they were even more firmly valued as *samting nating.* While there might be a market for insect bodies—connecting collectors with conservation places—Biangai were reluctant participants. Like the circulation of photos, samples, and other data, these were not the sorts of place-based relationships that Biangai had come to value. However, hosting visitors to their wildlife management area remained central to how they calculated the potential value of conservation.

SCIENTISTS, TOURISTS, AND SCIENTIFIC TOURISTS—OR JUST TOURISTS

Biangai generally refer to visitors as tourists, regardless of whether their work is formalized within scientific discourses or simply reflects personal and passionate enthusiasm for Papua New Guinea's unique flora and fauna. Conservation areas attract a variety of Western social categories, including ecotourists, scientists, and scientific tourists, which anthropologist Paige West describes as a social form with "very loosely organized research goals and no larger set of conversations in which they are taking part" (2008, 608). Instead, "self-fashioning and individual gain" (608) motivate their travels. When we view the problem from one end of the network that connects science types with their respective objects of study, we might rightly analyze the distinctions between varied social agents, as they reveal important power relations in an increasingly globalized political ecology. However, for Biangai all such visitors emulate what they understand as scientific practices, regardless of their positionality in Western discourse and society. Science types are rendered uniform as specimens of the same species, characterized by their values of "self-fashioning and individual gain." This is reflected in how Biangai engage with tourists—scientific or otherwise—and further highlights the reverse movement of generalizations in the production of values. Values are generalized as Biangai come to know all Western visitors as tourists who extract from the conservation area something of value to themselves. In this case, observational data and specimens are equated with photos and boxes checked off the life lists of birders, for example. While

Biangai did not see the value of insect specimens, they did value them as qualities of place that continued to attract possible social relations. Entomologists who visit are more important than those who collect from afar. In their experience, conservation worked when it strengthened relationships among people, opening the possibilities for "tourists" to become placepersons alongside Biangai. Maintaining these relationships, however, proved difficult.

During my research, visits to the Kuper Range Wildlife Management Area were in decline. Those who came varied in their Western career trajectories. Local guides spoke of being quite busy in the years preceding my research. The visitors I met represented a wide range of Western occupational niches: several amateur photographers (two specialized in insects), a weevil specialist, a moth specialist, a butterfly enthusiast, a Western scent company, students from the Bulolo Forestry College, as well as many ecotourists interested in the flora and the noninsect fauna of Papua New Guinea. They shared a number of similarities. Each had arranged the trip well in advance, organizing funding, visas, travel plans, and detailed itineraries for systematic data collection, opportunities for photography, or observation of specific kinds of nature. Planning is a hallmark of field research, and in repeated conversations visitors detailed their efforts to me, only to find that once on the ground their careful plans were destined to fall through. In many cases, the roads proved impassible after rain had washed out the only route. Another visitor was sent back to Wau after encountering an angry, "machete-wielding" Biangai man on the road, who was upset with Wau Ecology's failure to pay salaries to those working to maintain the research station.[9] Another had to rearrange his plans several times due to roads, mechanical problems, and weather before finally spending a rainy night outside the main camp, as the guesthouse was closed due to yet another dispute between a *solonarik* group and WEI. While their success in overcoming these and other impediments to their quest for knowledge might define their scientific identities as intrepid explorers of remote Papua New Guinea (consider the popular works by Jared Diamond, Bruce Beehler, and Tim Flannery), their persistence in the face of adversity revealed for Biangai the power of place to attract such persons.

In spite of such difficulties, or perhaps because of them, they all came away celebrating their experiences. "I have never seen so many insects at once," exclaimed a joyful photographer. The moth enthusiast described how he couldn't find time to sleep, as moth species were so plentiful he couldn't think of anything else. Likewise the weevilologist was simply overjoyed at finding so many wonderful examples, but he complained about the experience and was certain that were it easier, he would have more time with "his weevils." He spoke of the object of his study with pride, possessively noting how it would make for a very important and interesting dissertation. His weevils populated this area and were unique within the wider weevil world. In their accounts of field trips into the Kuper Range, values were at play. Each photograph, for example, was an attempt to capture something of the aesthetics of insects in place; it was a souvenir or trophy of the adventure, while it produced a highly marketable image in the stock photo trade (Frosh 2003). The capture of weevils—understanding their role in the wider ecosystem, placing their bodies in motion toward generalizations about a new species—also propelled the career of an ambitious PhD candidate. Even as they aided in this production of values for the far end of the chain, Biangai were making generalizations of their own.

Science types enter the Kuper Range, and through the work they perform, they transform how the community imagines the area. An experienced guide explained why tourists come: "They come for all kinds of work. Some come for ants, some come for birds, some for butterflies. Some come for all the little insects, or whatever. Some come for leaves, some come to collect all kinds of plants, or all kinds of little things. Some come for pythons [*he shivers*]. Some come just to roam around . . . wasting their time in the camp. Some come to go on top of the trees, by rope. . . . They are crazy." While interests among scientists vary, Biangai generalize the work as tourism and generalize the value of these visits as generating exchange opportunities. Much could be said about the uniqueness of Melanesian exchange values, and perhaps such narratives underlie the practices discussed here. But like the marketing of insects, the establishment of connections to tourists through places of conservation comes to replace all other values, "[rendering] them compatible with already established centers of [indigenous] calculation" (Latour 1999, 72). Opportunities for establishing and extending networks figure prominently in Biangai calculations as they participate in conservation. Conservation places make these connections possible, attracting not only money but also social relations between conservation placepersons and potentially powerful outsiders. But tourists often don't comply with these desires of reproducible sociality after their carefully planned trips are complete.

I spoke with Paipe, who accompanied the weevilologist on his trek, and he commented on how strange the trip had been. He and another experienced guide were exhausted by the work. It was not physically demanding, but they had stayed up late into the night. It had rained, and no one could sleep in the makeshift bush shelters. The weevil specialist could not speak

Tok Pisin, the common language of "his weevil's" human neighbors. Upon returning to the Wau Ecology Institute, the scientist could name the numerous species that he encountered and could discuss in detail the data that he collected, but he had only a vague idea of his Biangai guides' names, their relations to each other, or the relevant experience each possessed in biological fieldwork (which was quite extensive). While they were paid a little more than what was expected, nothing else was offered. There was plenty of rice, tea, and navy biscuits, but only a small can of sardines supplemented the evening meal. Favored tourists exchange small gifts or provide large feasts for those who share the camp. They offer contact information and promises to return (though few ever do). While visitors were often amazed at the size of the swarms revealed during these surveys, insects were but a medium through which Biangai sought to make connections, exchanges, and trades with their tourist guests. In this case, the scientist was only interested in his subspecies.

Taisan related a story about a couple from Canada from whom he had gained valuable insight into tourist world-making. While he was guiding them along the paths, the couple promised Taisan additional cash for each unique bird species they were able to see, record, and photograph. They were birders with long life lists, to which their trip to the Kuper Range would surely add novel species, giving them something to talk about with fellow birders back home in Canada. Taisan described how he had been leading them along a path to a cliffside with spectacular views, looking at the ground and not really paying attention until they offered more money for finding unique species. Then he was alert, searching for all the birds and other animals he knew, thus transforming how he worked. Taisan knew this area of forest from hunting and was as skilled at tracking animals as any Biangai, but now he was learning to pay attention in a new way to birds and animals that were not frequently hunted. He learned to attend to the photographic opportunities that tourists desire. Maybe he thought it would also further connect them to him, through the robustness of their exchange.

Identifying birds and other animals was something the young men in the village became experienced in, and many had trained to do such work. WEI set up formal training for ecotourist guides, and young men with these skills often called out the scientific names of birds and other species they heard or saw while working in the gardens. However, the ambitious weevil scientist, the other species specialists, and the multiple values that they attributed to their respective species did not connect them intimately with conservation places in the ways Biangai had hoped. For Elauru, it is the place that brings these diverse ecologies into conversation and conservation, while the many tourists come to value what they can take away from conservation places. Research trips by perfume scientists from Givaudan and students from the Bulolo Forestry College further highlight these dynamics.

SMELL MINING

Givaudan was one of the few major research expeditions to the Kuper Range Wildlife Management Area during 2001. Based in Switzerland, Givaudan is the world's largest flavor and fragrance company, producing chemical compounds for the food and perfume industries. In 2016, the company reported sales of 4,663,000,000 Swiss francs (at the time, approximately US$4,800,000,000). Composed of an international team of scientists, the expedition spent a week in the Kuper Range before moving on to other areas of Papua New Guinea and Southeast Asia. Because Givaudan spent about 9,000 kina (approximately US$3,000) on labor and repairs to the lab and dorm space built on the Kuper Range site at its founding, the visit was a major event for Elauru. Though most locals had little contact with them, everyone heard stories of the tubes and flasks, as well as the accounts of an older man speeding along bush paths with cane in hand, never slowing down and never falling behind. As the lead researcher overseeing the project, his eccentricities were the subject of much discussion. Speaking with an unfamiliar accent, he was a commanding presence at the camp, clearly a big man in the eyes of those attending the camp. The group was participating in what they called a ScentTrek, seeking out new smells in various conservation areas throughout the world. According to Givaudan's half-yearly report for 2001, "This year's first ScentTrek mission was carried out in Papua New Guinea. It was the first time that natural scents from this island were scientifically investigated. They are in the form of reconstitutions already available to Givaudan's perfumers" (Givaudan 2001).

This scientific investigation hired Biangai youth to lead them to aromatic plants, particularly those used in ceremonial performances. Some of these were brought to the campsite, while other fragrances were extracted from plants living in the forests without removing the flower. Using portable technology that allows researchers to monitor a plant's biorythm and collect odors during peak olfactory events, they were able to collect the scent molecules without constant monitoring of the plant. Givaudan trademarked this technology under the name ScentTrek. The name captures more than just its portability. "Scents" under this program of investigation move independently of the places in which they are embedded and independently of the plants as ultimate sources of marketable compounds. They are also abstracted from the bodily experience of the person who smells them. One researcher explained to me that they are not as interested in the composition that exists in the species of plant as in the components that can be extracted and remixed for the market. It is an extraordinary act of neoliberal dispossession. By breaking apart the floral elements in their labs, different compounds can be merged with other displaced chemicals, creating unique scents devoid of any connection to place. More importantly, their unique mobility allows them to file for trademarks and patents on the resulting formula. While they celebrate the practice as a sustainable use of the environment, the returns to the community are meager compared to Givaudan's financial gains in dispossessing smells from the environment. ScentTrek leaves no mark, sustainably extracting value from the forest, but establishes no relationships with persons or places. In this case, the value of this trek was quite momentary for the community but not for the company.

Givaudan also erases the particularity of place and persons in their presentation of research trips and marketing of individual elements. The relationships and practices of Elauru guides and others are hidden in the Givaudan marketing texts. Worse yet, the existence of people, including the Biangai, is hidden altogether. On its website Givaudan claims, "We also look to nature for creative clues on compositions and the discovery of new molecules to delight our customers. Our ScentTrek™ expeditions have taken scientists to many undiscovered and beautiful parts of the world: from the rainforest canopies of French Guyana and Gabon to the highlands of China and the cloud forests of Papua New Guinea to capture new scents that will provide inspiration for new fragrances or lead to new molecules" (Givaudan 2004). Such claims speak to proprietary rights and market the smells as unique commodities, unknown and sure to delight. Nature is the source of this creativity—not coevolution and cultivation through Biangai relations to place, not to mention Biangai enjoyment of the smells. For Givaudan, nature is beyond the horizon, "undiscovered and beautiful," in an undefined space without relations. "The cloud forests of Papua New Guinea" are the source of this supposedly creative endeavor.

When local communities are mentioned in subsequent interviews, they are removed from the present. As Swiss fragrance chemist Roman Kaiser stated in an interview, "Papua New Guinea was like an expedition in the 19th century. We flew in to some of the rainforest villages. They have a small airstrip because there are no roads. And most of them are situated on a hill. You land going uphill and take off going downhill. [It's] extremely exciting I can tell you, and from these villages you just walk and explore by hiking through the rainforests" (Brookes 2006). His description of the seventy-year-old Wau airstrip is quite accurate, but not only are there national highways connecting the villages to the coast, but Biangai and neighboring groups live in the same moment as the scientists, not the nineteenth century. Colonial and postcolonial history is erased along with the connections of the community to this new commodity and the scientists that "produced" it for the market.

The scientists who made the trek asked Elauru community members to assist by thinking about smells associated with their environment. Asked to help find "good-smelling" plants, like those used in ceremonies, Biangai knew exactly where to go. The best scents are cultivated in their garden spaces for future use. The scientists all spoke favorably of the conservation effort and talked about how their model aids in such efforts, creating "real value" out of the forest by discovering new scents and flavors for the market. Of course, these compounds weren't undiscovered by anyone, except for the market.

As with mining, resources are extracted, and profits are made elsewhere. There is a disjunction between the kind of work that is done by communities and the kind that the researchers imagine they have performed. The Biangai guides who have aided tourists over the years often gave me a name of a person, or a family or even a university, and ask that I find their "friend" who once stayed with them, gave them a knife, or just treated them with some respect. Such contacts are ultimately ways in which they try to extend their network of relations through place (see Fisher 1994), but for Givaudan's corporate strategy the Biangai don't exist (e.g., Brosius 1999). Instead, Elauru landowners are left to struggle over the meager returns from such ventures. While the places of conservation are folded into scientific markets, the extractive nature of these relationships avoids messy entanglements with local values, places, and persons. This collision of value would ultimately lead to a failure of the wildlife management area process.

IMAGINARY RASCALS

Residing in the household of those not considered primary landowners, I was privy to many flights of fancy like this one, recounted to me by some Biangai youth:

> Dressed in dark sunglasses, with the brims of their caps pulled low over their faces, four young men stood at the entrance to the conservation area. Professor Balun got out of his car and walked to the meet them on the road, where he waited for the group to

> explain themselves. The leader of the gang removed his glasses and stared at his adversary. Balun nodded, immediately understanding their purpose, and quickly handed over the check for three thousand dollars that was to be paid later to elders from the community. Taking the check and returning the dark glasses to his face, the leader of the group led them away, allowing the researchers to continue their work. They would spend one thousand kina to celebrate Independence Day, and the remainder would support them on their trip to Port Moresby.

The young men in my company on this particular September day had often complained to me about the way relatives managed the Kuper Range Wildlife Management Area. Such tall tales expressed these frustrations but rarely led to action. This time was no different, but it did serve to organize their thoughts as they prepared for another meeting regarding rights in the wildlife management area.

Regardless of the status of relations between WEI and Elauru, each year students and faculty from the Bulolo Forestry College spent a week at the Kuper Range Field Station training in the methods of data collection and species identification. In many ways, it was exactly the sort of relationship that community members desired. It was an ongoing relationship, as Bulolo was a short forty-minute bus ride from Wau. On trips to Bulolo with people from the community, we often ran into people from the college. My friends and I would chat and chew betel nut with staff and students from the college as we hid from the heat of the day. Biangai liked the forestry college's annual trip because it represented an ideal relationship through place. In early September 2001, the annual visit served to highlight tensions among landowners, engaging imaginations and necessitating a meeting. Professor Balun and others were expected to attend and to provide compensation for this particular outing directly to the community. As relations with WEI had temporarily soured, community members took control of arranging the college's annual visit. A letter was sent by a college-educated elder informing Professor Balun of the division between the community and WEI, and asking that payment for the annual practicum be made to the village. The elder explained to me that for too long WEI had *kaikai mani bilong ples* (eaten the village's money), so now he wanted to use the payment to purchase a cow for a holiday feast. Instead of just one family "eating money," everyone would eat cow.

On September 11, 2001, community members living in Wau and those with claims from other villages arrived in Elauru for a meeting about the forestry college's payment.[10] I had witnessed meetings of this scale in Winima several times. Disputes over land in communities often result in fights, and Elauru's meeting was sure to be contentious given the discussions that preceded it.

Small informal gatherings were held throughout the day. Talk of rights, land, and wasted money was heard from the various groups. I found myself situated in the household of my adopted family, who had voiced the complaint in the letter to Balun. From their perspective, the lands on which the field station buildings sit are partly theirs. Others supported them but always with the complaint that the current management had failed to equitably distribute benefits. Many hoped that a change in village-level management would effect changes in the how the area was managed, giving others the opportunity to work with tourists. One man argued that he and his family had provided a large parcel of land that tourists and researchers use, but his family had never received any compensation and were denied the right to work with tourists. Instead, only one kin group received rent, which they claimed as the caretakers of the area on which the research station rests. He ended his speech by noting that tourists don't come to take photos of a house.

The leaders of the project would correctly argue that when the topic of conservation first came up in 1992 village leaders uniformly agreed that the area where the field station was to be built was clearly under their control. Those now disputing these claims only held secondary rights, as they were related through a female ancestor (recalling distinctions made between hunting paths and village gardens, discussed in chapter 2 and 3). This was determined in a number of meetings where others lacked any interest at all in the area. It was not until there were regular visits from tourists, especially a *National Geographic* film crew, that interest was piqued. Furthermore, under the alternative development schemes established in association with the wildlife management area in 1997, the family managing the areas was given the guesthouse as their project. Others had projects too (beekeeping, recycling, a trade store, and a food bar in town selling cooked foods from the village), but each one had failed, surviving only in name.

The meeting was long and contentious. Without a resolution, Balun and others returned to the conservation area to continue their training, instructing the community to resolve the dispute before they received any payments. During the community discussion that followed, a fight broke out and one young man was seriously hurt. No agreement was reached. In the days afterward, those involved in the fighting expressed a good deal of shame by hiding in their houses or visiting relatives in town and other villages.

Places evoke passions and relations that are hard to disentangle. In the case of the Kuper Range Wildlife Management Area, multiple families can claim rights, as it is a node in a wider network of place-based relations. The effort to designate a single family as primary rights holders monetized the value of the past and made conservation difficult to maintain.

I asked my adoptive father in Elauru why this was never taken to the local land court. If they felt they had rights, surely this would be a way to resolve the conflict. He reminded me of a point that Philip (my host father in Winima) had made during a recent church meeting. It is best to work it out in the village, because once it goes to court, the community is changed forever. Even in dispute, places remain important for maintaining social relationships.

• • •

In seeking inherent values in nature, Westerners achieve "an immediate sense of ontological security and permanence" (Harvey 1996, 157). The international trade in insects as collectibles and the visit of "tourists," marketed as part of conservation, appeal to such ontological needs. But the problem is not only in the end product, encased in glass, bottled as abstracted scents, or stored in scientific collections framed away from the environment. The danger is in the network of relations that connects the visible marketable product to its ecological niche—to its place. While we can lament the insect's loss of life, its sacrifice is a celebration of values, as we are told on the small cards affixed to the back of the framed collections of butterflies in gift shops and scientific journals. Conservationists highlight such values, arguing that these meaningful relations can be easily transferred to threatened ecosystems. However, this value is not uniformly shared and experienced along the commodity chain—the network of relations that connects the sign to its referent.

Values, like knowledge, are phenomena that move and change along what Latour calls a "reversible chain of transformations," losing some properties and gaining others as they circulate and encounter distinct human subjectivities. While Latour's perspective remains grounded in the position of the scientific experts, what interests me here are the dialogical encounters between the creation of scientific phenomena and Biangai understandings of the value of place generated in this process. There exists a useful process of generalizing knowledge about specimens and samples as they move away from the places where they were collected, but a reverse generalization is at work here, because the scientists do not act alone. Biangai make generalizations about the values created through conservation practices. While species traits, habitat, behavior, and broader relations for flora and fauna are generalizable from their environmental complexities, so too are the practices and personalities of such specimens as conservationists, scientists, tourists, buyers, collectors, sellers, and photographers.

In this locally positioned perspective on the movements of specimens, visitors with highly varied skills training and practices are generalized as inhabiting a single social form, the tourist, while the objects they set in motion are disentangled from place. Biangai hoped that these objects would remain forever tied to the conservation area as they traveled, and that "tourists" would remain committed to places, returning and renewing their relationship in the same way as Biangai renew their connections to the garden. They hoped that circulation of conservation objects and the establishment of conservation places would create reproducible social relationships with potentially powerful outsiders. Neither proved to be true. This does not mean that the practices of varied forms of scientific and touristic activities should be halted, but there is a need to critically engage the values generated by their practices.

What is gained in places like the Canadian court discussed at the start of this chapter is a rich play of values, a debate about meanings and relations. We realize that the insect bodies are very much bound with placepersons in ways that are hard to disentangle. And why should they be separated? The image of Biangai children playing with butterflies and beetles—tethering them to village spaces and their own fragile hands—should remind us of the power of play in generating knowledge and values. Scientific practices in conservation areas, which encourage relations with local resource owners, need to conceptualize the interactions as more than an exchange of economic significance. Both conservation places and communities are fragile in contemporary political ecologies. As the tethers that held them together in the Kuper Range came apart in 2005, it became apparent that the local value of maintaining a conservation area simply lacked strength. In other words, their generalized experiences of scientific research did little to maintain local investments in conservation practices. The place, I was told, was tearing them apart, so they reclaimed the place, with its nuanced connections to families and histories, ending the Kuper Range Wildlife Management Area.

I don't fully condone the actions of Mr. Deslisle, but from a distance his exchange with Danielle Hauro seems much more involved and complicated than those experienced by Biangai in support of their Kuper Range Wildlife

Management Area. Instead of simple monetary exchanges, scientific research should be about sharing passions, appreciations, and the knowledge generalized from the fieldwork experience with those who live in the area and know the land. Such exchanges should have the goal of transforming insects and the like from *samting nating* to *samting moa* (something more), complicating the values experienced in conservation places in the same way that they are complicated elsewhere.

CHAPTER FIVE

Becoming a Mining Community

In July 2014, Philip and I sat in the dining room of his house. Its woven bamboo flooring and walls were worn and broken, and the house shook precariously as we moved around. It was the house that he had built and moved into in January 2000, and which I have visited over many years as he hosted me in the village of Winima. We talked about how he would tear it down and build a new one before I returned in 2015. Winima villagers had long anticipated that the mining company would help build new houses for the community or that they would be so well off that they would build new homes themselves. Having seen other mining communities in Papua New Guinea, they expected this and talked about it during my first visit in 1998 and on every subsequent trip. In the end, only a handful were able to accomplish this.

As we talked, Philip noted that every company is different, and many had tried to establish a mining operation at Hidden Valley at the head of the Bulolo and Watut Rivers. Conzinc Rio Tinto of Australia, Australian Goldfields, Aurora Gold Limited, and Abelle Limited all failed to realize the mine's potential before Harmony Gold Mining Company (South Africa) took over as gold skyrocketed in price, from a 2005 average of US$444.74 per ounce to a peak average of US$1,668.98 per ounce in 2012.[1] Having established Morobe Mining Joint Venture with Newcrest Mining Limited, Harmony and Newcrest were able to begin production in 2009 on two large open pits—one at Hidden Valley and the other nearby at Hamata. But the promise of a transformed village never eventuated, at least not in the ways Winima villagers anticipated and desired. Some changes were apparent. Children from the community were better educated than those from neighboring villages, more people worked for the mine and its supporting contractors, many migrated to urban areas, and a number of families started small businesses with contracts at the mine (transportation, site rehabilitation, and so on). However, life in the community was slow to realize the sort of transformative potential that mining had seemingly promised (see also Jacka 2015). Philip insightfully attributed this to changes in the company. He observed, "Each company has a way about it. But it is not just the company. Management too, and the employees are all different." Companies have not just a different corporate structure, but different corporate polices, different mediators of cooperate policies, different objectives (exploration, construction, production), different prices of gold, different labor conditions for workers, different promises, and different ideas about how to negotiate with Papua New Guineans (see Golub 2014 for a detailed account of such negotiations). Each variation translates what we call mining and what we call neoliberalism in significant ways (e.g., Welker 2014), and each requires that we attend to its unique process of assembling the relationship between places and persons.

The organization of mining and agricultural labor in Winima reflects growing inequities in the development of a mining economy and demonstrates that the market increasingly mediates social relations. However, Biangai living in both Winima and Elauru retain important attachments to place; neoliberalism has not dominated this essential aspect of being Biangai. Place and social relations are affected by mining labor and the financial opportunities that gold offers Biangai communities. Neoliberal projects promise to make new subjects: workers, citizens, and an aspiring middle class. Biangai long for such experiences because of the perceived political and economic benefits, and because gold had already proven its power to transform. However, as with failed conservation initiatives, disputes around mineral rights and growing economic disparities among participants are implicated in the "creative destruction" of social relations (Harvey 2005). What hope Biangai retain remains with the ongoing production of places and the social relations they hold together.

The relationship between Biangai and gold is realized in the alluvial deposits that line the streambeds around the communities and in large reserves of gold managed by corporations. Whereas small-scale mining seemingly fits with community life, industrial mining is much more transformative. Small-scale mining is treated like gardening, while the corporate practices of industrial mining inform community life. While these changes frustrate some ambitions, such as Philip's desire for better housing, they also facilitate creative and novel engagements between Biangai and place.

PANNING FOR GOLD

Like Orokaiva of Oro Province in Papua New Guinea, Biangai view modernity as constructed around perceived differences between themselves and white expatriates and colonists (Bashkow 2006). Western development and modernity are linked through the practices associated with Europeans (timekeeping, mobility, income). However, Biangai also extend this conceptualization to the material world via gold as part of that from which their modernity is made. Perhaps this is the result of gold's value in projecting and sustaining the hope of development and the relative ease (compared to coffee, scents extracted by Givaudan, specimen samples, tourism, and scientific research) through which it is made into modern wealth. Biangai understandings of gold are both literally and imaginatively folded into placemaking. Alluvial mining still evokes a relational network connecting the living and dead and grounding experiences of place. Ultimately, gold is a gift transacted between Biangai and the land.

In the 1920s and 30s, Wau was the site of a series of gold rushes that marked the creeks feeding into the Bulolo with the productive efforts of artisanal Australian and European miners (Healy 1967).[2] Continuous with these efforts, alluvial deposits were scooped up from the riverbed by multistoried California-style dredges operated by corporate interests, while richer veins of gold were exposed through tunneling (Waterhouse 2010). By the end of the 1920s, five thousand miners and attendant laborers inhabited Biangai hunting grounds in what was at the time one of the largest gold rushes in the Southern Hemisphere. In the 1930s, Wau emerged as a major center in the colony. For Biangai villagers living on the edge of these developments, places were forever transformed. Rivers that had been the location of spiritual forces were physically remade through the shifting of soils and cutting of timber that lined their banks. And the valley, once populated by tall klinkii and hoop pines, reemerged with plantations of coffee, modern colonial architecture, European vegetables, and cattle.

But overlooked in this standard narrative of colonial expansion are Biangai views of why non-Biangai came to the valley. For Biangai, their hunting and gardening practices are not the first images that form the valley floor. Instead, they look toward a deeper history. Elders in both Winima and Elauru explained that Wau was immediately recognized as important by early European explorers. In fact, they argue, it was Wau that they were looking for. A common narrative tells of a historic flood filling the valley (evident in the geological record), matching biblical accounts of Noah. For many Biangai, this fact is part of why whites came to Wau, returning to biblical places. One spoke of Captain James Cook "discovering" Port Moresby in the 1770s and claimed that Cook visited Wau before returning home to describe "what sort of man lives here, as well as what kind of ground, gold, timber, all of the things that are needed for work." While his account was a bit vague on the details (and contrary to current historical understandings), it echoes the sentiments of another who described how whites had come to realize that Wau was the source of many things. In a long discussion of languages of Papua New Guinea, Kausa, a Biangai man from Elauru, detailed the linguistic similarities that, from his perspective, diverge from Biangai vocabulary and flow outward across space. He suggested that this is evidence that these groups originated in Wau. Kausa also traced the dispersal of klinkii and hoop pines (*Araucaria hunsteinii* and *Araucaria cunninghamii*, respectively) from Wau through neighboring valleys. Likewise, he surmised that the gold which miners continue to carry away must have an ultimate source. As water, gravel, and silt flow downhill, Kausa argued that gold too flows down the valley from the mountains of the Kuper Range around his own village and into the rest of the nation. The gold now being mined is ultimately derived from Biangai places. An elder once shared with him a dream about a truck that kept on carrying what seemed to be an unlimited supply of gold. The truck would dump its cargo along the Bulolo for others to take "like it was coming from a factory," but "they carry it from where? The truck starts and carries the gold from where exactly?" His rhetorical questions speak to a claim that gold mined through industrial or artisanal mining ultimately comes from Biangai lands.

The Upper Bulolo, Kausa and others argue, is the source of many things. According to these mytho-historical narratives, missionaries and explorers knew this too and were simply following the path to its source. Such narratives speak of a great historical unfolding that will slowly bring Wau back to prominence. Furthermore, Biangai narratives about gold reveal the belief that there is a great deal of wealth buried in the upper reaches of the valley. Kausa explained, "The gold on top has not revealed itself. The gold hides." Gold's agentive qualities here result not so much from its reliance on or independence from human actors as its relationship to human practices and placepersons. Failure to recognize such relationships is the reason that gold is hard to find. Once Biangai relationships through place are properly recognized, the gold will reveal itself again. Tied to their relations with the other-than-human world, places of gold, like yam gardens, are a source of hope. Patience, Kausa says, will unearth much for the next generation.

OCCASIONAL MINERS

Given the relatively long exposure to colonial and postcolonial mining, one of the more curious aspects of Biangai socioeconomic practices is the uneven investment in small-scale artisanal mining. While there is a belief that the bulk of the mineral resources remains hidden as Biangai sort through their connections to place, gold continues to be mined throughout the valley. But given their claims to the land and their prominence in and around Wau, Winima and Elauru community members remain only occasional small-scale miners.

In comparison to Watut speakers who work on what Biangai still claim to be Biangai land,[3] Elauru and Winima villagers are a minor presence on the alluvial goldfields of the Bulolo Valley. Only a fraction are full-time small-scale miners. While some Winima had rights to work known deposits through family in Kwembu village (closer to the town of Wau), few were working these during my research, as mining royalties made the hard work of alluvial mining less appealing. During minor rushes, they would join Kwembu-based relatives (in the same patterns as if they were planting a garden). Among Elauru's residents, one family held a government lease that they worked continuously throughout the year. While they maintained a well-appointed house in the village, they spent most of their time in Wau. For them, it was full-time work. Furthermore, the head of the house had been a leader in various Biangai associations and held some sway in regional politics. Living in town facilitated these activities, and mining helped fund his lifestyle in Wau. In contrast, small-scale miners in Winima and Elauru do short stints once or twice a year to pay for needed items or to cover funerary events, travel, and other extra expenses. It is work that is more frequently pursued by young men, though most had some experience either working for others (of European or Papua New Guinea ancestry) or on their own claim at some point in the past. Sam explained his mining practice: "Two weeks in Wau, I go for somewhere around two weeks. It's not like it's full-time. Say I am in the village, coffee season is finished, and I feel that I am still in need of something, so I go for two weeks to Wau, three weeks perhaps. Get the money . . . divide it up . . . return to Elauru."

At the time of the interview, Sam had been married for ten years. The eldest son of a village leader, with eleven years of formal education in the coastal city of Lae, he was accustomed to certain comforts that required money. He joked that he'd like to buy a generator and a television to watch videos in the village. He explained that working all day in the garden to return home and fall asleep while watching videos would be wonderful. It was one thing he missed from his time in town. In addition to this bit of wishful thinking, two of his children attended the local community school, and he had to meet his obligations to pay for their school fees, school supplies, and clothing. While his children didn't really know what he called the "life of money," he'd like to provide some of the comforts of urban life for them. Gold was one way to do that without having to leave the village too often. Village life, he explained, was better than living in town, because when money was short one could always find food and a place to sleep. Full-time mining didn't appeal to him for that reason. Occasional mining helped fill the gap and maintained his family's connections to place.

Kawna had a similar explanation of why he worked the alluvial deposits of distant family closer to Wau: "I go to work in the goldfields in order to find the good life for my family," but "I only work for a short time." Furthermore, "if I need something, like my family needs something, okay, I have to find a way to get some money." To pay school fees, afford rice and religious obligations, or help with funerary expenses, he needed to find extra money on occasion through gold. Though older than Sam, Kawna was educated and experienced at urban life. He was among the first generation of Elauru villagers to attend high school outside of the area. In contrast to Sam, Kawna's father had worked in the goldfields to help pay his school fees, though he too was never a full-time miner. His children were older as well, and one was attending high school at the time, with additional fees for room and board adding to the higher costs. Money was a main concern, but moving to town or working the goldfields full-time was not an option, as he associated the good life with living in the village. He explained, "I think of my kids and that is why I work," because "I have to show them a good way to live." Panning for gold was associated in Kawna's view with money. When I asked him, "What is money?" he responded that money was food, "because when I have money, I have food." However, he elaborated that it was not real food, but a kind of "strength." Like gardening, gold mining was one way he sought to find strength.

Kausa offered a different story. As a youth, he stayed with his father in Wau. At first, his father worked for the local medical clinic, but later he worked as a small-scale gold miner on a leasehold owned by an expatriate from Australia. Kausa lived with his father while he was at the clinic, but as he got older and his father moved on to working the streambeds, he was sent back to the village. In remembering this, Kausa attributed the change to both his own youthful transgressions and the difficulty his father had looking after him once he started working gold. In his own words, Kausa was becoming a *bikhet* and a *raskal* ("stubborn" and "a rascal"). He was sent back to the village so that he could learn to value community life and know his father's lands. The goldfields lacked these values. During the course of my research, Kausa made two trips to work alluvial deposits near Wandumi; he was never gone for very long and went only because he was invited. He was formally trained as a Lutheran evangelist and supported a wife and two children, and working in the goldfields helped him to fund his mission activities on behalf of the Lutheran Church.

Working gold competes with a variety of other practices as well, and Biangai attitudes toward work significantly affect how they view artisanal mining and other labor outside of the village. In interviews on different kinds of work, small-scale artisanal mining, building a house, and planting a yam garden were ranked as the three most difficult tasks. The latter two are also central to Biangai identity in general and masculine identity in particular. But in contrast to conservation commodities, gold is increasingly linked to contemporary Biangai identity. In many ways, it is comparable to yam production. Yams are typically described in Biangai origin stories as emerging from the body of the first man as his gift of sustenance to future generations (see Martin 1992), but some now speculate that gold is the true gift. Furthermore, the work of planting yams is seasonal, occurring toward the end of the coffee season (though increasingly delayed by the growing work of harvesting larger crops). Like gold, yams require only occasional work.

More significantly, yams and gold are sensitive to human social relations linking Biangai persons to places. Yam gardens typically involve subsets of kinship groups with shared rights to the area being planted. To ignore a family member with rights to the area is to invite supernatural or magical interference in the productivity of the garden. Biangai extend such understandings to wider environmental relations, as the *imeng* or "soulstuff" (Maschio 1994) of the dead is dispersed across an active landscape. But these agencies are a part of their living relations, as they animate the landscape their living bodies once traversed. For the land to be productive, Biangai must have recognized rights and good relations to those with existing rights, both living and dead. Hunters, for example, follow specific paths to specific locations. Hunting outside of those areas minimally ensures a failed hunt but could result in bewilderment and even death. Such relations among persons, places, and things defined the productive practice of Biangai in precolonial contexts and continue to model expectations for gold.

K was from a different province. He married into Winima after meeting his spouse in town. While he was recognized as having rights to work his wife's land, he remained non-Biangai. Late one day while hunting, he discovered a small amount of gold in an area long associated with mineral exploration. K was thinking about returning the next day as he fell asleep, and he dreamed that he would not be able to find the gold again, that spirits would bespell him or the gold would vanish entirely (see Moretti 2006 on gold and dreams). Following the insights of his dream, instead of returning on his own, he informed his in-laws. Shortly after I left in 2002, the gold was rediscovered and a small rush occurred. However, the kinship group only invited members of their extended family to work the area. They quickly exhausted the site before returning to their regular tasks. Similar stories of small rushes are quite common, with rights to work the deposit organized along the same lines as gardening in general. Disputed areas often go unattended until some agreement can be reached. This was the case in 2015, when gold was found along the border between Werewere and Elauru villages. With competing claims, the gold remained untouched until those with relations to both villages quietly worked the small deposit.

Working the land, tilling the soil for a garden, or sifting through it for gold is indicative of how Biangai "dwell" in their environment (Ingold 2002), and thus revealing of Biangai naturecultures and em*placed* values. In their discussion of gold mining, Biangai often describe its place using the Tok Pisin word *graun* (or *ngaibilak* in Biangai), which can be translated as "earth," "dirt," "ground," "land," or even "the entire world." In describing the work associated with gold mining, GK notes how one comes to know the *graun*:

> Working gold, I've done it. I feel working gold is, gold is found deep inside the *graun*. And there are many kinds of *graun*. In one kind of *graun*, it will be two meters and at that layer is the gold. In some *graun*, it can be fifteen meters, ten meters, seven, eight, nine meters inside. So I pan. The first work to find gold is that . . . I pan. Then I find out about this *skin graun* [surface layer]. . . . It could be a hole or a landslide, or some other kind of deposit, but I pan for gold.

The *skin graun* is better understood through turning the soil over and over in one's pan, looking for signs of change in the layers. Throughout the task is one of removing soil, removing stones, breaking them, and sifting. Getting to know the *graun* necessitates working with it, attending to its color and texture while looking for the small flakes of gold. Once found at the surface, digging deeper can reveal a richer deposit. GK went on to explain that in working with a sluice box, the task is the same, though a flow of water is employed to facilitate breaking up larger amounts of *graun*. For the men interviewed on this topic, breaking and removing defined the work, distinguishing it from other labor, where growth and renewal were thematically deployed. Though the work requires knowledge and skill, as GK suggests, gold already exists in the *graun*.

For those familiar with small-scale gold mining, GK's description of the work is not unique. However, gold's emplacement in *graun* or *ngabelak* is significant for understanding how Biangai view the gold itself. Like planting yams, the soil is turned with the goal of making it produce something. The *graun* of yams is made productive through this effort (removing bits of root and rocks to give the yam room to grow). Gold requires a similar labor and also requires proper sociality among those who work the claim. As I sat around the fire at night with a group of young, mostly unmarried men, our discussion turned from Hidden Valley Gold Mine toward the possibility of gold in nearby creeks. One observed that gold is a kind of "free money." Others agreed, noting that when found, the gold is yours. While yams are grown and nurtured like children, gold only needs to be unearthed. It is a gift from the land to those who have the right to be there.

In contemporary Biangai mythopoesis, gold is sometimes conceptualized as the body of the first man. His sisters had been instructed to tend to the grave and watch for the emergence of something that would take care of Biangai long into the future. In the older versions of the story (see Martin 1992), the shoots of yams hold the promise for the Biangai future, but this was a future without development, money, colonization, time, whites. In the new future, it must be gold that will be produced as a gift from an ancient garden. But productivity, as for all gardens, is based on sociality, and Biangai in Elauru and Winima viewed gold as hiding until they had worked out the proper relationships. This is why gold hides along the Upper Bulolo River and why the Hidden Valley mine has faced so many difficulties. A series of accidents, including one fatality in 2015, were seen as an ongoing sign of these concerns.

At the most basic level, Biangai living in Elauru and Winima are limited to occasional mining by the lack of bankable reserves on the surfaces of their land. Geologists have told Elauru villagers that it is not technically feasible to mine their true resource, as it is too deep to merit the effort. Thus, Biangai in Elauru work at the invitation of family living closer to town and in the occasional small gold rush around the village. For example, relatives in Wandumi invited Kausa and others to come work a landslide for a week. A pastor and two of his parishioners did similar labor on the land of a fellow Pentecostal church member to raise money for a conference in Lae. But this limited access does not entirely define their relationship with the gold. Important to Elauru and Winima's conceptualization is their location at the headwaters of the Bulolo River Valley. Their conceptualization of the headwaters as the source has been confirmed for Winima by the development of Hidden Valley Gold Mine at the head of the Watut and Bulolo Rivers. For Elauru, their close association with an important men's house and the presence of the first man, whose body is said to be buried there, are both central to their claim that ultimately all gold originates in the headwater communities. Even if it is now buried so deep as to make it difficult for them to obtain, they assert, with some patience and certainty, that events and the land will unfold favorably for future generations.

On the morning of the account that begins this book, Kawna failed to find gold. While his efforts were not successful, they maintained linkages to past events, practices, and mytho-historical narrative. His careful gaze into the pan marked the space with possibility, while acknowledging the reality of gold's uneven distribution across the valley. Throughout their encounter with colonial and postcolonial mining, Biangai have been intermittent miners. For those living at the upper end of the valley, this reflects the limited amount of accessible gold and historical alienation of lands closer to town. But this is not the interpretation they offer. Biangai space flows out from the Bulolo Valley, folding and unfolding events that include past, present, and future mining. Elauru and Winima villagers approach their current position with patience and hope. For Elauru it is just a matter of time, and for Winima Hidden Valley Gold Mine holds such promise.

INDUSTRIAL MINING: "IT'S NOT THE SAME AS BEFORE"

Over the course of ninety-five years, the Biangai have participated in mining as indentured laborers under the Native Labour Ordinance (1922), free laborers, free wage laborers, independent miners, and more recently, coparticipants and employees in Hidden Valley Gold Mine.[4] Now, when Biangai interact with the gold-mining community, members are more likely to be relatives, neighbors, and other nationals. Under postcolonial mining, Biangai, along with neighboring Watut and Buang speakers, line the streambeds heaping shovels of earth into makeshift sluice boxes, serve as physical laborers for larger claims and international companies, and increasingly work in the offices alongside expatriate managers. As beneficiaries and employees, they stand to gain more from the exploitation of their lands than did their grandparents' generation. But they also gain new insights into global markets, urban life, and distinctions defined through class and ethnicity. When villagers, and even company employees, quip, "It's not the same as before," they evoke these historical changes in the political economy of mining, mirroring my host father's observations that differences persist in the way mining is practiced and experienced by different actors. This is true of the two contexts for Biangai: in the alluvial fields of the Upper Bulolo and in the large open pits of Hidden Valley and the neighboring Hamata mines.

Discovered in 1928, the ore body at Hidden Valley remained too deep and too remote for development until recently. Since 1983, exploration rights (EL677 and EL497) for Hidden Valley have changed hands numerous times, with five companies (Conzinc Rio Tinto of Australia [CRA], Australian Goldfields [AGF], Renison Goldfields Consolidated [RGC], Aurora Gold PNG, and Abelle Limited) attempting to develop parts of the area. Eventually Harmony Gold took over, forming a joint venture with Newcrest Mining, and they started production in 2009 under the name Morobe Mining Joint Venture. In 2016, after declines in their return on investment, Newcrest sold its shares back to Harmony. Today, Harmony Gold is the sole operator of the site (table 5.1).

For residents of landowner and non-landowner villages, these companies vary not only in their ability to develop the mine but also in their relationship with the communities around Wau. The variation is important, as each emphasized different aspects of capital, its potential and pitfalls. Villagers compare corporate generosity in the provisioning of goods and services, the variability of work opportunities, and overall ability to provide development. More often than not, they have concluded that the company is not the same as before. Initially, Conzinc Rio Tinto of Australia was quite generous, perhaps reflecting concerns raised by civil unrest at the Bougainville Cooper Mine (Filer 1990). As the price of gold declined, different corporate actors were more reticent to provide the extra benefits and close relationships that the community had come to expect. Once the memorandum of agreement was signed and production was underway, gifts and favors from the company almost completely ceased. Now, community relations officers refer Biangai appeals for "support" to the landowner association, the local government, and Biangai businesses. By contract, these entities are supposed to be more directly engaged in the community during production, replacing the mining company's role as a generous negotiator during exploration. In 2016, Harmony closed down the office in Wau in favor of maintaining offices in Lae and at the mine sites, moving their staff away from visits by local leaders and concerned community members. Community leaders complained about this change, as it disentangled the company from the places where people reside. Their complaints were ignored.

Over the course of multiple research trips I have observed the opportunities that industrial mining presents at different stages of development. Mineral exploration at Hidden Valley began in 1984 and lasted through 2005. During this time various companies were involved in the management of the mine site. In fact, over its history some seven companies have attempted to develop and operate a mine at Hidden Valley (table 5.1). Once the company identifies a resource, feasibility is assessed, a mining license is obtained from the Papua New Guinean government, and an environmental assessment is conducted. Once approved, construction begins. At Hidden Valley this included roadwork as well as building the processing plant, tailings dam, a conveyor system, offices, employee facilities, and so on. The productive phase began in 2009, with the first gold bar realized in June 2009. Table 5.1 details mining company relations with community members and workers

TABLE 5.1 Phases of mining and corporate involvement at Hidden Valley

PHASE	COMPANY	DATE CORPORATE INVOLVEMENT BEGINS
Discovery of gold in the area	William Chapman	Early 1928
Exploration to initial feasibility studies	Conzinc Rio Tinto of Australia (partnering with Placer Pacific Ltd. and Carpenter Pacific Resources at various stages)	March 1984
	Australian Gold Fields NL	February 1997
	Aurora Gold Limited and CDC Financial Services (Mauritius) (purchased Australian Gold Fields assets)	September 1998
	Abelle Limited (merged with Aurora)	January 2003
Feasibility to Construction	Harmony (purchased controlling interest in Abelle)	May 2003
	Harmony-Newcrest joint venture (70%–30%)	April 2008
Production	Harmony-Newcrest joint venture (50%–50%)	July 2009
	Harmony (sole ownership after purchase of Newcrest's stake in the mine)	September 2016

during the periods of exploration, filled as it is with hopeful uncertainty, and production, characterized by routine and growing income inequality as well as urban migration.

Historically, the development of large-scale industrial mines is associated with a drastic transformation of community life.[5] Both outmigration and harsh work conditions negatively affect the predominately male population (e.g., Walsh 2012), women enter the hidden economy as prostitutes (e.g., Hammar 1998; Wardlow 2006), and in some cases the entire community around mining is undermined (e.g., Richards 1939). Postcolonialism has brought about a great deal of variation in the experiences of these communities and the kinds of relations that working for the mine entails at home (see Banks and Ballard 2003; Jacka 2018).

In Northern Brazil, for example, late twentieth-century mine workers moved somewhat systematically between an emphasis on the individual and the social—a contrast expressed idiomatically as *manso* and *brabo*. *Manso* (literally, "tame") references experienced miners who have been socialized through dwelling in the *garimpos* (mining area). "People enter as raw individualists, are transformed by the collective nature of *garirmpagem* [gold-digger], and come to realize that in such a difficult and hostile environment cooperation and unity are absolutely essential for survival" (Cleary 1990, 134). In contrast, *brabo* (literally, "brutal") imagines the *garimpos* as "wild and anarchic" (134), where individuals are driven by greed. The frames of sociality are not exclusive to specific regions but emerge from the negotiations and contestations around gold.

Such contrasts are also apparent in images of home and country along Africa's copperbelt communities. "Localist" workers, who still maintain homes in rural villages, return for intermittent visits to the communities, much like the Biangai mine workers. For them, home is formed around the "bonds of kinship" and uterine images such as the place where one's umbilical cord is buried (Ferguson 1997, 140–41). Town, however, is a prostitute, immoral and lacking such filiations. The localists value their connections to home and exert a good deal of effort to maintain them. In contrast, the "cosmopolitan" middle class celebrates the values of rural life but wishes to remain in town. This contrast signifies that "dislocation is more often a partial and conditional state of affairs, an uncertain predicament that entails neither a clear sense of membership in one's community of origin, nor an uncomplicated conviction of having left it behind" (153). As urban Biangai continue to work their coffee gardens, occasionally returning to plant yams, they reflect such cosmopolitan sensibilities. But they also reveal a relationship with places that is often ignored in the literature.

These unambiguously defined mining relations in Brazil and Africa, moving between individualistic *and* kinship-based systems, contrast with the case of Bolivia, where an indigenous cultural system supports expression of agency and a kind of resistance. Instead of conflict and contrast between systems, Biangai practices suggest a creative local engagement with the work of mining, as each place-based kinship group seeks to benefit from the relations that mining has brought into being. Even as mobility increases, connections to place remain powerful if not troubled. These relationships intensify as the mine shifts from exploration to production.

EXPLORATION

At the sound of an engine, easily identifiable as that of the Toyota Land Cruiser, children poured out of their houses, running to where the road entered the village, yelling at the top of their lungs, "AURORA I KAM!! AURORA I KAM!!" It was a common refrain in 2001, as Aurora Gold employees traveled to Winima to purchase produce for use at the mine site; carry announcements and food for Christmas, funerals, and other occasions; transport equipment for company-sponsored sporting events and courses in the community; as well as pick up and drop off villagers going to and from town on company business. These visits marked how most villagers interacted with the company on a weekly basis.

While many young men had spent some time working at the mine site, and most of the village had visited the offices in Wau with a personal request or a complaint, it was these weekly encounters with drivers, community relations officers, and other company personnel (most of whom are Biangai themselves) that shaped the contours of community-company interactions. It was through changes in these encounters that community members felt the effects of the mineral economy to which they sought affiliation. The quality of these interactions suggests that Biangai are realizing a greater degree of tension in how they experience sociality and place.

CONTRACTED LABOR

Many Biangai worked for Aurora Gold during my initial long-term fieldwork between 2000 and 2002, and most of the men from the village of Winima had worked at least part-time for the company at some point. During the exploration stage of the mine, two Winima men worked in the Community Relations Office, two worked in security, one cooked in the mess at the mine site, and one woman purchased garden produce for the company kitchen.

Others performed tasks as welders, drivers, and manual labor. Biangai from the village of Kwembu were equally represented in the Morobe Goldfields offices, and Biangai men and women from other villages filled positions as full-time drivers, security officers, and office staff. With one community relations staff person based in the village (as a liaison and voice for the company) and others in place at every point of contact between the company and the community, including drivers, security, and office staff, Winima villagers were limited in their contact with expatriates and Papua New Guineans from other regions.

Working with the company provides much for the individual employees involved and their families, including income, on-the-job training, educational opportunities, and a sense of involvement in the development project. It affords them the opportunity to pursue the coveted middle class lifestyle (Gewertz and Errington 1999). However, during exploration, these employees/landowners became more than intermediaries for company policy. In many instances, they absorbed much of the criticism about company decisions that might be directed more effectively at Morobe Goldfields, as their actions were seen as evoking place-based relationships more than corporate responsibilities. Biangai working for Morobe Goldfields were viewed by many villagers as being "one-sided" and as representing only the limited interests of their kin group over that of the whole community.

In contemporary Papua New Guinea, the integration of class and experience is partial; with the assurance of customary rights in land and resources, the boundaries between owners and workers are not as distinct (Gewertz and Errington 1999, 10–13). Since neither the state nor capitalists are able to alienate mass quantities of land without landowner permission, class distinctions are limited to status distinctions.[6] As with the Porgera mine, the Biangai "think of work as belonging to those who had immediate authority over them, and not the mining company" (Imbun 2000, 142). Since those in such positions were Biangai, kinship and relationality through place mitigated distinctions of status and insulated the company from direct critique. Thus, Biangai community relations officers were not seen as acting as individual employees, but rather in terms of their relationship to the community. This would continue to be a problem as the mine went into production, but was exaggerated during the exploration phase because full-time work was limited to small groups of short-term contract employees. Whereas the company demands individual accountability and job performance, the Biangai expect sociality and commitment to kinship through place.

During Hidden Valley's exploration phase, the company hired villagers on short-term contracts to clear helicopter pads and work sites. The

Madang-based drilling company, United Pacific Drilling (PNG) Ltd., contracted others for exploratory work. Throughout the fieldwork, these three- to six-week positions were the cause of internal fights, trips to the village by Biangai community relations officers, and debates at village meetings. Because two of the community relations officers were from *solorik* groups with strong claims to the area around Hidden Valley, others complained that contract employment opportunities favored their families over others.

Tensions were elevated during an afternoon of educational videos brought to the village by one of the community relations officers. An audience of young males was to be shown videos on HIV/AIDS, pregnancy, and with a little luck, a Jet Li movie. They had set up the television and generator in the Winima Community Hall, built years before by Conzinc Rio Tinto of Australia. It stood in the center of the old village site as a testament to the company's generosity. Over the years, it had suffered quite a beating as the target of community anger over the mine. Its dented corrugated iron walls, broken door, and graffiti testified to the mine's positive and negative influences on village life. In previous disputes, anger was often directed at village property and not the company. On this occasion, as the crowd of youth watched educational videos, an argument broke out. Upset about being ignored in the recent round of contract hiring, a few members of the Igulu *solonarik* began arguing with others present. One member of the group dumped the spare generator fuel on the floor and lit it with a match. While the fire spread across the plywood, a village leader and a pastor ran to the scene, only to be confronted by an angry Igulu youth brandishing an axe. No one was hurt, and the fire only scarred the surface of the floor, but tempers ran high and community leaders called a meeting early the next morning.

During fieldwork, I alternated weeks between Winima and Elauru villages in order to compare the different development directions they were pursuing. While this provided me with a much-needed comparative perspective, it often resulted in startling moments, such as the one that confronted me on this particular return trip to Winima. Unaware of the argument the previous night, I came into the village in the middle of a heated discussion about the events. Leaders stood in the middle of the main road, while other men and women stood along the sides, lined up according to their *solorik* affiliations. Much of the blame for the dispute was leveled against the community relations officers for being one-sided, for picking only their family and allies, for acting in terms of a limited relationality, and for not carrying out their responsibilities as community members. They called for them to resign their positions.

The community relations officers were subject to divergent expectations. On the one hand, the company insisted upon their adherence to policy, to objectivity, and to the value of individual responsibility. They were workers. On the other hand, the community was divided between those who held the same expectations, in order to ensure their equal access to benefits, and those who expected employees of the company to act in terms of place-based obligations. As a result, any given action was subject to multiple interpretations, as well as the employee's ability to manipulate the expectations of all parties.

Members of Igulu argued that they too had the right to contract work, and they believed that in the most recent round of opportunities they had been intentionally overlooked. The unofficial policy was to distribute employment opportunities across the village and across families. This policy reflected the company's interest in widely distributing benefits across all sectors of society (perhaps with the hope of avoiding jealousy), but tensions remained from the ongoing dispute over which *solonarik* would go first in future decisions about Hidden Valley. The Igulu group stood firmly on the court decision awarding a 25 percent stake to Winima village and not to a kinship group. Further, it was argued that their ancestors also came from the land around the mine, and many could claim secondary if not primary rights in the area. The Igulu envisioned a larger network of place-based relationships. Paro and Kaigowe men and women argued that it was their land, their hunting paths and lodges, and their ancestors. This was especially true for those of Walaplang—a path between larger kin groups. For some of them, the Igulu had no right to complain if they did not work, because it was not their mine. Paro and Kaigowe sought to cut Igulu off, making different claims about land and ancestry, imagining different place-based relations with those who, through other paths, might be called kin.

Tempers were elevated, and a fistfight ensued as Paro and Kaigowe men charged into the center of the crowd. They were angry about the threats of violence against community leaders, angry about the claims of rights to work on what they perceived as their land, and angry that they would lose the right to work until this dispute was resolved. In the confusion that followed, I stood to the side among the women, who jeered for and against the melee, and two company employees (both Biangai) who watched, turning their heads from side to side in disappointment. They had seen this happen many times before.

The violence of that afternoon was not new, nor would it be last time the community was torn apart by conflict. The condition of the community hall testified to earlier disputes, and everyone had a story of broken promises, falsified genealogies, and threats made against them by others. What is of interest here is that the discussion surrounding the fight focused mostly on the community relations officers and not on the company and the broader political ecology of gold mining. In another dispute over employment, a man had threatened to set fire to a one of the company's Toyota Land Cruisers, whose symbolic value was not overlooked by the community. All sides warned him against this, arguing that this was no way to deal with "community problems." Again, when disputes arise, the company is not to be blamed.

Under the exploration agreement, hiring policy emphasized experience and a good record in order to ensure that the work was efficiently accomplished. A few Winima villagers would eventually develop a bad record, if they had not done so already. Furthermore, the Madang-based drilling company (contracted by Aurora Gold) made all of its decisions in Madang, faxing a list of which employees they wanted hired. Yet it was the Biangai employees that everyone questioned. Their location between relational commitments to place and the individual responsibility of company employees allowed for such questions.

The divisive character of company-community interactions could also be read in the organization of community spaces. The land of the village was reterritorialized by the threat of violence, as well as actual violence, as tensions continued. This occurred in terms of both the creation of physical barriers within the village and access to garden lands. Like most Biangai villages, a group of siblings established a row of houses in one section of the village, another group next to them, and so on. However, unlike any other village at the time (even those closest to town, where cattle and crime were more problematic), Winima was notable for its bamboo and barbed-wire fences.[7] These began to appear over the course of 2001 and intensified after a round of fighting. While publicly the fencing was attributed to the occasional cow escaping its enclosure, privately many confessed that the real reason was a fear of conflict and the ongoing disputes within the community. These fences did not surround kinship groups, as with precolonial village barricades, but enclosed individual households. Effectively, fences regulated movement in the community, defining and defending spaces in novel ways and structuring community sociality. Paths that used to wind among houses through the community were transformed, ending the chance meeting with neighbors. Even those who did not previously frequent such routes could read the visible signs on the village landscape. Those who put up the fences not only imagined restrictions in space but foresaw changes in relations and brought them into being.

But more importantly, Paro and Kaigowe villagers began to express greater reservations about secondary resource rights to mined places and, even more significantly, to question the primary and secondary rights granted through and to women. They emphasized certain aspects of how places inform relationships, while deemphasizing others. Like the wildlife management area in Elauru, some even offered the suggestion that rights for women were a fabrication, a product of gold exploration at Hidden Valley and the efforts of other Biangai to benefit unfairly. In effect, acknowledging cognatic claims through female ancestry would acknowledge the rights of a much larger group to the meager shares of compensation and services coming from the stalled exploratory work.

Further, as these claims were disputed, garden land too became contentious both for individuals seeking to establish a new garden and for children who might otherwise be designated to follow a mother's or grandmother's rights. One man explained to me that while he had rights through his mother to use a particularly nice garden area near his house, he was forced to continue to plant his gardens farther away, on his father's land. The area closest to the house remained off limits as others disputed his maternal claims to mining compensation. A family that could garden on land of one ancestor might also claim the hunting paths deeper in the forest belonging to that ancestor. By attempting to cut the network (M. Strathern 1996), Paro and Kagowe members sought to increase their control over benefits and mined places.

In this, they continued to act as placepersons by strategizing through the affiliations of kin and entanglement of place to foster stronger ties to the mine. They were not individualized landowners, securing their rights against all others. Exploration varied qualities of differential corporate generosity, and irregular work opportunities seemingly resulted in a more segmented identity. While such struggles characterized much of the tensions during the exploration phase, the shifting corporate entities that operated the mine increasingly relied on the court ruling to limit their involvement. As they moved toward construction and production, the company increasingly treated Winima as a singular entity.

PRODUCTION

Much of the tension in early employment derived from its limited and irregular nature. As the mine went into full construction and production, more than two thousand people were hired. The hiring priorities strongly favored men and women from the three landowning communities. According to the memorandum of agreement, job applicants were divided into five tiers:

Tier 1: Landowners
Tier 2: Bulolo District residents
Tier 3: Morobe Province residents
Tier 4: Papua New Guinea citizens
Tier 5: Non-citizens

This differential employment practice resulted in emerging inequalities between landowning and non-landowning communities, most notably in terms of urban migration and residence, small business participation, increased educational opportunities, and reduction in gardening.

While landowners are supposed to be given priority in hiring, there is some debate about how effective this policy has been (Burton 2013). Criticizing Morobe Mining's claim that 50 percent of landowners were employed by the mine (including contractors), Burton estimated that tier 1 only accounted for 13.6 percent using the data collected as part of his social mapping on behalf of the company. Given the flexible nature of kinship and residency, it would be difficult to pin down an exact percentage. From my household surveys in Winima, there was a great deal of variability each year, and changes occurred each month as employees were hired, fired, or simply quit. Prior to signing the memorandum of agreement, Winima villagers aggressively competed for short-term contract work at the mine. In 2011 every household had at least one person employed by the mine or by a mining contractor. Many had multiple members working at the mine. While the opportunity to work persisted, by 2014 many had left work. Work conditions and the simple experience of having to follow orders were the main reasons people gave for leaving on their own. However, some lost their positions after engaging in fights, breaking rules, or failing drug tests.[8] Thus, both the company and Burton could be correct at different points in time, though the 50 percent claim would be high in light of the data that I have available to me.

Biangai from non-landowner villages were often able to secure employment as well. While many resented being lumped in with the rest of tier 2, connections and skills made them ideal candidates. For example, fourteen men and one woman from Elauru worked at the mine in June 2016. Their complaints were the same as those from Winima, but they were more reluctant to leave their jobs. One employee who had many concerns about working for the company noted that he didn't have other options. In Winima, he explained, they have royalties and compensation payments to fall back on. With a twelfth-grade education, he didn't want to return to relying on coffee. Mining had entered what Simmel called "the calculating exactness of practical life" (1936, 196), requiring a certain level of commitment to the money-based urban economy. However, we can't view all employees monolithically (Welker 2014), and ultimately Biangai ideas of place and sociality provide spaces for respite from the painfully modern calculation of mineral labor.

Joe, my host father in Elauru, was one of the first from his village to be hired by the mine. In 2000, he took a job as a driver while his wife and children remained in the village. He was mainly charged with operating a Toyota Land Cruiser, transporting employees, landowners, and supplies between Wau, Lae, the mine site, and the landowner communities. During my research, he lived in town in a rented house (2000–2005), his older brother's house (2005–13), and his own compound in the Biangai settlement in town (2013–present). The cost of urban life was Joe's main complaint, but maintaining a house in Wau allowed him to look after children from his extended family as they attended the area high school. Education was important to him, and working allowed him to support his children and others, saving on boarding at the school and providing income for food and school fees. When the mine shifted to construction in 2007, he quit his job. He complained that the 6:00 a.m. to 6:00 p.m. work schedule was too much. He was also tired of working "for someone else." For many years after retiring, he maintained a beat-up dump truck, driving it on contract for locals as well as for several companies.

Joe's concerns with education reflect shifting commitments to a changing economy. At the same time, he maintained his ties to the village, spending holidays and breaks there when he worked for the company. In recent years, he has returned to the community for extended periods to help manage his family's coffee, as well as for important funerary events, community meetings, and holidays. Ultimately, he offers a complex image of commitment to the social reproduction of community life. Gardens and family matter greatly to him, but so does education and the employment opportunities it affords. Working full-time for the mine differently positions Biangai into novel centers of calculation, where "qualitative values" can be reduced to "quantitative terms" (Simmel 1936, 196). But local ideals still matter (e.g., Gudeman 1986).

This was apparent in the many interviews I conducted with mining employees on break. Returning home from a two-week stint at the mine,

Yamu sat with me around the fire one night. Throughout my research, I have often asked returning mine workers how things are going. It is a casual question, one that I ask friends and family all the time in the United States. And like my American friends, Biangai have common expressions to talk about work that characterize the labor, their relationships, and their attitudes toward the job. Yamu thought about my question and then responded, "It's okay, but it's not okay too much." I laughed and said that others had told me the same thing. He nodded and explained, "Yeah, we are all in prison up there." A 2017 Facebook post by a Winima man who works at the mine echoed these sentiments, noting that he was going "back to jail." Other employees responded to his post in agreement, noting that prison might be better. Sykes has shown how common idioms demonstrate "widely felt experience and publicly expressed concern" (2001, 8) as well as reflecting culturally constructed understandings. In this case, Biangai workers were concerned with rules and the impact that work was having on their bodies. Prison, with its mixed social settings, imposed schedules, rules, and diets, was an apt comparison (see Reed 2003 on Papua New Guinea prisons).

RULES: *STRIK TUMAS*

The men and women who express dissatisfaction with work typically complain about rules. The work, they would say, is not hard, but the company is "too strict." Each day they work twelve-hour shifts from six to six. During this time they are heavily regulated and watched, but many expected this. It is the off hours that substantiate their complaints. After work, they have a short time to clean up and eat. If they are late, the company dining hall will be closed. Then they have a short time to relax before they must be in their rooms. One employee explained that lights must be off at 9:00 p.m. if you are working day shift. No music in your room, no bathroom trips, no phone calls—just sleep. One night, his neighbor was playing music after hours. Security came to turn it off and remove him from the dorm. This was cause for dismissal.

Workers told many stories regarding thefts from the company and the ways they would break company rules. Many talked about ways to sneak in banned items like betel nut, a stimulant favored throughout the country. The mine had banned its use due to the red spit that is produced as the meat of the betel nut mixes with the lime powder with which it is prepared. Biangai found many ways to slip this past security checkpoints. Others joked about petty theft, tricks to get out of working "too much," and fantasies of stealing bricks of gold. "It is ultimately our gold," they would quip.

Other common complaints involved random searches, strict security checks at the gate, and the disposal of worn out clothing. Employees during exploration were given old clothing, and it had become a common item of exchange with friends and family, connecting the place of mining to the wider community of placepersons. Starting with the construction phase, this policy changed. Boots that were deemed not fit for work were thrown away. "They are still good," employees would say. It was a minor complaint, but it was telling of the different approaches to regulating behavior and shifts in corporate policy that troubled Winima villagers in particular. Biangai workers found ways around this restriction. This was often done in the context of giving, like the worker who took extra clothing for some relatives who had wandered up to the mine. Dressing them in company uniforms, he escorted them to the mess, fed them, and showed them around the camp in broad daylight. For the worker, it was his way of showing his kin that he cared about them and that he could take care of them. It was a bit of bravado, but he acknowledged obligations to kin while displaying his own status as an employee. At the same time, he risked this status. In addition to various safety training sessions and other workshops, these regulations shaped worker lives, even as they found ways around them.

As they lived on site (a full day's walk from the village), the rules extended into their sleeping patterns and off-work behavior. Working for two weeks with a weeklong break, effectively they lived two-thirds of their lives in camp. There were only two shifts, and each employee was assigned to either A, B, or C crew (table 5.2). Each crew worked one week from 6:00 a.m. until 6:00 p.m. and then another week from 6:00 p.m. until 6:00 a.m. On the last day at the end of their night shift, contracted buses sent them home. Most were dropped off in Bulolo and found their own way to their houses and villages by nightfall. Sleep during their time at the camp was governed by a calculus of curfews and dining schedules. The schedule of the mine required that they consolidate sleep in ways that were not typical of the everyday in the

TABLE 5.2 Mine work schedule for different crews

	A CREW	B CREW	C CREW
Week 1	6:00 a.m.–6:00 p.m.	Break	6:00 p.m.–6:00 a.m.
Week 2	6:00 p.m.–6:00 a.m.	6:00 a.m.–6:00 p.m.	Break
Week 3	Break	6:00 p.m.–6:00 a.m.	6:00 a.m.–6:00 p.m.

community. Instead, they revolved around neoliberal institutions that simultaneously normalize sleep and destabilize a sense of normalcy by alternating night and day (cf. Wolf-Meyer 2012).

Workers were told to take it easy on breaks, not to work too hard and to stay healthy. They were expected to rest and return to sleeping at night so that they were ready for the day shift upon returning. Many who came to the village on break adhered to this guidance. They played cards, hunted, participated in community events, and worked a little in the garden and around the house before going back. Quite often, employees on break would stay in town, with many taking up full-time residence there even as their families remained in the community. Coffee and yams continued to demand their attention during the harvesting and planting times. For many, the schedule was disruptive of communal labor that would be performed in the garden, religious commitments to their churches, and the everyday sociality of community life. Crews often separated siblings and close relatives, the very people whom one might rely upon for shared labor. If your siblings were on different shifts, you would never run into them in the village and only in passing at the camp, as your everyday routines depended on your crew assignment. Instead of planting yams as a family, you would plant yours on your break and they would plant during their breaks. This disruption of village social life has long-term implications for how people know a place. Recall that yam gardens are shared endeavors; family groups work together to plant gardens. While most continued to plant their own yams on breaks, in a few instances families planted on behalf of absent workers. Not working together leads to different kinds of relations to place, different knowledge about garden histories, and different kinds of relations to other-than-human worlds, emphasizing individual labor and obligations to one's immediate household. That work structures the everyday lives of those who labor is certainly apparent in both mining and gardening. But the values of place and relationships to the products are quite different. However, there are certainly signs of garden labor being influenced by mining practices.

Early one morning, around 6:00 a.m., I joined a Winima man as he and his extended family went to work together in their coffee garden. He was on his weeklong break from the mine and was eager to see the state of his coffee. His wife had promised him that there was much work to be done. Having finished a small breakfast of plantain and tea, we headed up the Bulolo River to his coffee garden. At the time, he had a total of 1,057 trees planted in a single garden, relinquishing other rights to his brother. This garden was on land gained from his father. By concentrating his groves in a single area, he had simplified the work of tending to coffee, making the trips home less arduous (his subsistence gardens were planted nearby), but this also limited his potential for expansion. On this day, his family, his neighbor's family (who shared a common ancestor on his father's side), an Elauru youth (who shared a common ancestor), a Watut man, and I joined him in the garden.

There was much to do, and we began work picking the ripe coffee cherries around 8:00 a.m. The work was tedious, and we moved from branch to branch and tree to tree in a silence that was occasionally broken by short conversation, jokes, and discussions of the work. At noon we ate lunch when he checked his watch, and we left shortly after 6:00 p.m., again following his watch. His wife complained about this strict schedule, noting that when he was working at the mine, she did not keep such a schedule. With years of working for Hidden Valley in his background, he responded that when he worked for the mine they followed the clock. "Why," he asked, "should it be any different here?"

Adherence to European punctuality takes on a moral significance. For Kragur villagers on Kairuru Island off the north coast of East Sepik Province, a failure to adhere to time, to keep time, or to manage their time better was understood by community members to mean that they were inferior to Europeans (M. Smith 1994, 217). While Biangai do not attribute European success to supernatural powers (at least at the present moment), they do use relative adherence to temporal regimes to differentiate kinds and ways of working the land. Winima Biangai commonly joke when they see each other busy in their gardens or around their houses that they are truly "six to six," marking the temporal regime of the mine. However, the mine discourages hard work on breaks, providing a reason for some to relinquish garden obligations to kin. Increasingly, mine workers noted their obligations to the company and the need to rest for work. As they experience work through specific temporal regimes, their relationships with the material world are likewise moderated, becoming farmer and worker in a novel set of ecological relations.

WORKER BODIES

Worker bodies were also shaped by changes in diet and the less arduous physical demands of the labor. While sore backs and strained muscles were common alongside lost fingers and torn tendons, much of the labor was sedentary: driving a vehicle, operating equipment, or sitting in an office. When discussing labor at the mine, the hard part, mine workers consistently said, was working for someone else and meeting his or her expectations. The work itself was not hard. But, as a result of the less physical nature of mining production, the shifts in diet were even more apparent in the bodies of employees. Many gained weight while working, developing noticeably increased fat and decreased fitness. The ready availability of fatty meats, grains, and starches on a daily basis, combined with a tendency to spend the first few days of a break drinking copious amounts of South Pacific Lager, was worn on the bodies of many mine workers. At least three had stopped coming back to the village because they found walking to the community too arduous. The also feared having to walk back to town at the end of their breaks, a task that takes most members of the community three hours but would take them much longer.

During the El Niño drought of 2016, many employees were laid off due to water shortages at the mine site. As a result, the number of employees living in the village at the same time afforded the opportunity for a series of formal and informal focus groups. During one such interview I sat with a group of men and women who worked in the company kitchen (run by Niolam Catering Services).[9] I asked them about the foods they prepared each day. Without hesitation one said, "It's okay, but not so great." Others agreed, but the complaint had more to do with taste than overall quality. The menus were planned well in advance, they explained, and they were set by the company to follow specific regional themes, in addition to regular offerings of chicken, pork, and beef with vegetables and rice. Each night had a theme oriented around a regional cuisine from around the world. Often, these reflected some segment of the international workforce in technical and management positions at the camp. But the themes bothered the local workers; they had an Australian night, a British night, a Philippines night, pizza night, a Chinese night, and even an American night. But where, they asked, was the night for Papua New Guinea?

Their concerns about national theme nights were not about the work of cooking different foods; they actually liked learning the recipes. Instead, it was about the taste. For Biangai, taste was one of the common purveyors of difference, delineating both styles of cooking and where it was produced (Halvaksz 2013b). Food grown in their soils simply tasted better than that grown elsewhere. While some enjoyed the variety as it made them feel cosmopolitan, others found themselves eating copious amounts of the same thing every day because the other options were not appealing. The regional tastes were not popular with many other local employees as well. Except for one employee who especially looked forward to pizza night, many preferred to eat foods more typical of Papua New Guinea (including sweet potatoes, greens, rice, meats, and fish). Taste is also a product of where the food was produced. "Everything is from Australia," one long-time employee complained. "Even the tea, it's not regular tea . . . it's too strong. If you don't drink it while it is boiling hot, it will cook your mouth." Other foods were seen as spicy, with workers commonly noting that they fought in your mouth. Even locally available foods (like early companies had relied on) were replaced with imported meats and ingredients. At the end of our discussion, one long-term employee complained, "It just doesn't taste like Wau." Dietary preferences thus reinforced differences between local employees and others as place figures into taste (Halvaksz 2008b). But the larger point was that the regimes of work with their regulated temporalities, tastes, and expectations were generally not favored by Biangai workers. For these reasons, many chose to leave their jobs or seek employment outside of the Hidden Valley mine.

RELATIONSHIPS BETWEEN WORK AND PLACE

In contrast to the prior phases, production at Hidden Valley increased the number of Winima villagers who could be employed in mining. As a result, workers' lives were highly structured by labor practices, their bodies transformed by new diets and habitual movements. How they related to place was also altered. While many remained committed to the village and the garden, an increasing number acquired or rented land in urban areas. The combination of living some distance from the community and accommodating the crew work schedule strained their ability to participate in the social reproduction of their communities. However, many Biangai continued to return to their coffee and their gardens. By 2016, a number had quit altogether and returned to the community, or found work in the small businesses that supported mining. While the former clearly reflected a commitment to place, the latter allowed for more flexibility in movement. Small business employees had time to return home during yam and coffee season, as well as for community events and holidays. For now, their sense of place remains tied to this. The creative destruction of social life is alarming but not absolute. In the Biangai worldview, hope resides in those spaces—hope that housing will change, that working conditions will improve, that the mining company will better acknowledge landowners as partners. Next year will be better.

Biangai employment has led to a degree of ambiguity, as those hired are caught between being individuals who sell their labor to the company and being members of the community obligated to various networks of kin and place. Their obligations, however, are not without problems, as tensions persist. These tensions would continue, with violent and deadly consequences. At least one person has died as a result of place-based disputes within the community, and ultimately as a result of ongoing disagreements about mining benefits. The deceased was buried along the main road, where he was killed in a fight. As with precolonial burial, where bones were distributed across landscape associated with the deceased, his bones mark the land. While this tragedy clearly divided the community and provoked ongoing discussions of mining's impact on their lives, most eventually returned to working in their gardens or to their jobs at the mine. Unlike Elauru, they can't just stop the mine. In spite of growing instability, the continued commitment to place—its gardens, tastes, and sociality—suggests that we can't understand mining as a singular event; mining shifts from exploration, to production, to closure as companies come and go.

While the tensions in the community might prompt indigenous mobilization against mining (e.g., Kirsch 2006; Hyndman 1994) and the development of a regional identity (Jorgensen 1996) in the long run, mining is not a static process. Instead, we must view gold mines historically, as affiliations and relations emerge and diverge around the place-based political, economic, and ecological relations of production. Employment is one location of these slippages, but so are the practices associated with small-scale mining, gardens, and conservation. Mines themselves are limited in their temporal presence by the bankable reserve and, many Biangai hope, their impact on places. As tensions are worked out, relations through land are likewise transformed. Here, the emergence of fences and transformed sociality of workers suggest debates over the limits of Biangai relations through place. Others fear that they will lose connection to places altogether, and see a future that is increasingly fenced in.

CHAPTER SIX

Whose Closure?

WHAT might mine closure and closure planning mean for Biangai communities? Mined spaces are landscapes of a particular kind, ones that the best practices of closure will find hard to erase. Mines not only provide minerals but mark time and space with their appearances. The closing of the conservation area provides a case in point, as it continues to evoke other potentialities. Biangai from Elauru imagine new tourist sites, conservation programs, and businesses linked to ecological research. Attention to the aesthetics of a development project's *appearances* and Oceanic concepts of time and space is useful in considering Biangai placepersons in the context of frontier capitalism and local ideas about relationships through places.

Past development successes (and failures) are like shadows cast over time. They appear as future possibilities, blurring the linear "progression" of history. They speak to temporalities that combine attention to time as present-processes (Gell 1992) or that view time as "a symbolic process continually being produced in everyday practices" (116) and Pacific epistemologies of time (Māhina 2010). As Samoan scholar Epeli Hau'Ofa notes, "That the past is ahead, in front of us, is a conception of time that helps us retain our memories and to be aware of its presents" (2008, 67). The rich histories of mining and its material affects are not forgotten, as they remain alive in the present. They point to time as experience, subject to local frameworks for understanding its relation to place, discourses, and practices. Attention to "the action frame of reference and the shallow time of everyday life" (Gell 1992, 314) prompts a focus on appearances that instill aesthetic qualities of the past into the shape of everyday life. Reflecting on Tongan spatiotemporal relations (Ka'ili 2017), we can see how the history of mining has served to "mark the spaces" of production with the great wealth that mining can and does generate. They are, like the rhythms of *tā-vā*, marking space with each pit, each corporate actor, each sluice box moving mountains of dirt from the land. Gold's uneven distribution orchestrates the production of aspirational spaces of development. Mines become places of hope, where modern neoliberal subjectivities can be realized. With closure, these pasts don't go away, but rather remain as ancestors of future development. Examination of these appearances can reveal the place-based aesthetics of time under Oceanic ideas of *tā-vā* and the rubric "folded space-time" (Serres 1968, 1982; Serres and Latour 1995). These are the aesthetics of gardens, hunting paths, and other sites of great historical meaning, as well as flakes of gold that erupt through the layers of earth's folds.

Time and space can be understood as working together to create an aesthetic, such as beautiful dances or artistry, in which temporal events are marked by the steps of the dance, the stroke of a brush (Ka'ili 2017). My argument here is not that mining is beautiful or that Moanan philosophy and art should be compared to resource extraction. Instead, I'm articulating how Biangai think with and about mining. As emergent neoliberal subjects, they remain committed to the work of making meaningful relations. The appearances of these ideas, objects, or practices reveal the role of desire in shaping such spatiotemporalities.[1]

Mines have a physical presence—a mnemonic that folds the past into the present—but their aesthetic qualities are as significant as their multitemporality, especially as they exist in relationship to placepersons. For instance, mines transform the landscape, but these transformations remain subject to multiple interpretations. How mining inscribes landscapes has often been described as loss (e.g., Rumsey and Weiner 2004). But mines can inscribe attractive qualities as well. Appearances acknowledge the unrelenting aesthetic presence of contemporary and historic practices as they continue to provoke and entice. However, appearances are more general in location and less intentional in their creation. Part memory, part practice, and part discourse, appearances resist the irreversibility of time; they are folds in time and space—a turbulence in the system (Serres 1982, 71–83). They can be unexpected, ghostly, appealing, or repulsive but are always powerful and persuasive and emplaced by the world-making practices of a mining community.

These are not new perspectives: Biangai places have always been invested with aesthetic qualities that attract and repel gardeners and spirits, animals and humans, forming and divesting social relationships. Biangai anticipate the perpetuation of mining not only on the basis of the geological presence of gold but also on the basis of the landscape's social history. The Wau-Bulolo Valley has been mined since the 1920s, and like the ancestral gardeners and hunters of the past, companies and miners have adorned the landscape with the techniques and technologies of their trade: the deep pits, deforested spaces, eroded mountainsides, discarded equipment, the buildings and structures of a township, a skilled alluvial workforce, not to mention the wealth that was gained in the process. These practices are as much a part of how the land appears and its potentialities as those associated with the yam or coffee garden.

Mineral exploration and extraction can have an attractive hold on a community, while leaving repulsive sites of exploitation and environmental degradation that inform a landscape of a different kind.[2] Biangai share Moanan or Oceanic insights about time in their engagement with the appearances of mining. Their space-times are folded and gathered together but much more grounded in the contours of the land and the geological folds of past activities. It is this grounding that continues to afford them hope that development will eventually be fruitful.

NEW GUINEA GOLDFIELDS' APPEARANCES

In contrast to small-scale alluvial mining, New Guinea Goldfields (NGG) sought the deep golden veins that stretch under much of the Upper Bulolo around the township of Wau. NGG created the sort of spectacle that certainly attracted attention and fostered certain expectations about potential, much like the exploratory phase at Hidden Valley. The company appeared successful, if only for the moment. It established itself at Edie Creek, Koranga, and Golden Ridges in 1928, accumulating small leases and expanding these claims in 1930 with the acquisition of Guinea Gold's leases. NGG installed telephone lines, established the first roads between their mining leases, provided power for the township, invested in office and residential buildings, opened a freezer plant and bakery, funded the famous expeditions of Mick and Dan Leahy that opened up the Highlands region to colonial control, and explored the possibility of land routes to the coast. At the start of the 1930s the company was perceived, rightly or wrongly, as the major player in the Morobe Goldfields.

With a nominal starting capital of £5,250,000, the largest mineral investment in Australasia at the time, the company controlled much of the discourse about mining along the Bulolo.[3] NGG eventually used £4,500,000, with much of its shares going to acquire leases that were exchanged for smallholder rights. In doing so, it created not only a large claim but also a large body of knowledgeable investors eager to see a return. At the shareholder meeting on March 27, 1933, Mr. Daly, an investor from Sydney, complained, "I put money into this concern on the reports of men who have been there, and what have we got? Nothing. Unless you people wake up and make some move the day will come when you will be told that there is nothing there, and that your four millions of money have gone to the wind" (*Pacific Islands Monthly* 1933b).

With respect to such international investment capital, Anna Tsing argues, "Investors are looking for the appearance of success. They cannot afford to find out if the product is solid; by then their chances for profit will be gone" (2005, 75). But NGG's investors were fully aware of the field's potential and were certain that "the product" was "solid." They fretted, instead, over the delays and the growing perception that any dividend paid would be spread over such a large number of shares. In NGG's case, "economic performance [was] conjured dramatically" (Tsing 2005, 75) but not unrealistically. The company had not overestimated the gold reserves (cf. Tsing 2005), but its ability to quickly turn a profit along this resource frontier proved to be less than magical.

The editors of the regional news magazine *Pacific Islands Monthly* wrote most critically, calling for restructuring on numerous occasions (*Pacific Islands Monthly* 1931, 1932a, 1932b, 1933a), and at one point noting that NGG's position was "not so pleasing" (1933c, 6). While NGG's executives held that a dividend would eventually be possible (*Pacific Islands Monthly* 1933b), shareholders and the media questioned how they could be rewarded for their investment and how the company could expect to raise the additional capital needed to finish developments at Wau. NGG spent wildly in its early days, expanding both its leases and the number of individuals with an interest in the company's performance. However, even more capital and thus even more shares would need to be issued to bring the operations into production. Too much time had been wasted in exploration and construction. For investors, the appearance of failure is especially troublesome, and NGG was increasingly seen in this light.

NGG struggled over the next twenty years, facing three strikes, a slow turn to profitability, the destruction of their facilities during World War II, and a relatively difficult postwar reconstruction.[4] Obviously, these temporary cessations in mineral extraction were not closures, but they serve to highlight the less than certain nature of mineral extraction. Along the resource frontier, mines are often held together by relatively unstable social and financial relations and not simply the identification of bankable reserves. Like the Hidden Valley mine today, NGG was very much tied to appearances. The enormous start-up capital, the rapid extension of leases, and the heavy investment in developing the town did suggest a highly successful future. The company moved with great velocity, attracting and deploying capital in shaping Wau's spatiotemporal contours (Harvey 1982). Within the expatriate and international finance communities, the appearance of success required a concrete return on investment. The only alternative would be to close down or divest.

The current project at Hidden Valley is no different. Struggling to establish sufficient reserves, the mine changed hands numerous times before a partnership between Harmony and Newcrest brought it into production. But repeated financial and technical difficulties have resulted in repeated rumors of closure, speculations that were amplified with Newcrest divesting in 2018 and the visits of other foreign investors. Like NGG, Hidden Valley painted a positive image for its investors even as the mine experienced repeated difficulties. In contrast to carefully cultivated investor perceptions, these "appearances" played out quite differently among the Biangai. Such differences highlight how greatly stakeholder perspectives on closure vary, informed by different perspectives on mined landscapes.

Biangai and other communities around Wau do not exist in early accounts of NGG. They are erased from the frontier or naturalized into its landscapes (see chapter 2). Biangai and nearby Watut (Blackwood 1978; Burton 2003) speakers are only mentioned as threats to the development of the region (Booth 1929; Clune 1952; Idriess 1933). For most Biangai, memories of working among the miners follow WWII, and only a few recall prewar interactions focused on produce markets and trips through Wau. Colonial policy at the time would have encouraged recruitment from other areas. Biangai gardens and gardeners were used to grow produce in support of mining activity (Chinnery 1998), and Biangai were employed as carriers by alluvial miners (Booth 1929; O'Neil 1979). Furthermore, during the interwar period, nearby communities were involved in providing logs for mining companies. A Biangai man who had been raised in Wandumi (the closet Biangai village to Wau) explained to me how his father was employed cutting timber that ended up as support posts deep within the Edie Creek mine. In spite of his present age, he remains fascinated by these early operations. During one of our conversations, he described how his father took him and some of his family to the opening of the deep mine shaft at Edie Creek. They stared with a good deal of trepidation at the cave. Its darkness was lit up every few meters, with the timber posts supporting the roof like a house. Though short on benefits, Biangai were not completely out of the picture. But without a financial interest, the work of the company could appear quite attractive. This era continues to exert a hold on the Biangai imagination.

Over the past thirty-five years, the early activities of NGG have been the subject of much discussion. Biangai complain that their wealth was stolen.

In many ways, they are caught up in the "appearance of success," where NGG's *past* "economic performance is [still] conjured dramatically" (Tsing 2005, 75). However, it is not simply the extraction of gold and corporate profits that garner local attention but the persuasive power of the land's social history. Spaces mined by NGG, or any of the small-scale miners for that matter, retain more than the company's name and inform more than a historical narrative. While a similar case could be made for the Wau Ecology Institute and conservation, the practices of mineral extraction remain more salient in local perceptions and valuations of the landscape. These historical relations are folded into placepersons who are shaped by the potential for new exploration, new investment, and new relationships with outsiders, but this time, on better terms with Biangai, who are most connected to the land.

For the company, these were difficult and hardly profitable times, but they still transformed the valley floor into a bustling township. In Biangai history, the Bulolo Valley is intimately linked with the death of a great ancestor and the production of the first yam, conjured from his flesh. More recently, it was the location of conflicts between Biangai and neighboring Watut speakers. Throughout, it was a forested hunting ground, peppered with many hunting lodges of Biangai cognatic kin groupings or *solonarik* (Burton 1996a; Mitio 1981). In creating a township, not only did NGG reinscribe mythological and lived landscapes as wild spaces to be transformed through its version of frontier modernity, but it adorned this place with the structures and practices of mineral development. The appearances of these pasts inform Biangai ideas of what development should look like, and these appearances continue to attract them to mining.

Biangai and other area residents are very much caught up in the appearances of places. Of course, NGG was eventually successful. Despite investor concerns and protestations to the contrary, it did extract wealth from the Upper Bulolo and its tributaries. And this wealth was distributed throughout the expatriate community through wages, subcontracts, the purchase of goods, and so on, as well as to the wider Australian and colonial government through tax revenues, some of which in turn funded government services for the territory. For Biangai, gold mining conjured development most dramatically. NGG's appearance is incorporated into the landscape as a potential quality of present and future projects (as with Kirsch 2004). Like a yam garden, mined spaces serve as a shared mnemonic in a Biangai landscape, evoking historic and potential relationships. But it is the aesthetic qualities of mineral extraction (the attractive constructions of town life, mining equipment, roads, and so on) that are most powerful.

Such transformations of place highlight tensions between different valuations of land. In the creation of the postcolonial mining township of Paiam, the Ipili rejected the commodification of land by multinational developers. They insisted on treating the exchange as a gift with long-term obligations, performing a ceremonial exchange that "reproject[ed] their relationship with their ancestors and land into a space of capitalist development" (Jacka 2001, 8). Paiam, like Wau, reflects the appearances of mining, attractive and appealing to local sensibilities in ways that frontier capital barely appreciates.

Biangai at the time of NGG were in no position to make the sorts of claims made by the Ipili, as Biangai land was deemed "waste and vacant" by the colonial administration. While defining what development looked like, NGG and the administration denied the local population a formal role in mineral production. In the absence of the company, Biangai are very much divided on how to proceed. Many want to make their claim in court. Biangai are locally known for going to court. Expatriates and other Papua New Guineans in Wau object to the Biangai penchant for using the courts to seek compensation, often complaining to me that Biangai would rather go to court than start a business. They are contrasted with the neighboring Anga speakers, the Watut, who are seen as industrious and hardworking. Biangai have successfully used the courts to win claims over Wau's lands, in various development projects, and against neighboring linguistic groups, and they have defended their court victories through many subsequent appeals. For the Biangai I know, going to court is both an effort to seek compensation and an effort to reproject "their relations with ancestors and land into a space of capitalist development" (Jacka 2001, 8). Court is viewed as a ceremonial performance, not unlike community meetings over disputed lands where one's claims to place (*kasi mek*, literally "a road") and those of one's ancestors are narrated and evaluated and placepersonhood is performed. Unlike companies, courts too fold time and space, with the power to deny "closure" and redress wrongs. It is in the courts that Biangai temporality can be expressed and appreciated in globalized networks (consider Kirsch 2004, 2014). One might conclude that the court has become a modern-day rite of renewal.

For example, in 1980 Biangai landowners claimed 17 million kina was due to them from the various parties that benefited from their gold (*Post-Courier* 1980). It is a complaint that surfaces every few years. One Winima man, who has repeatedly sought to obtain my help in this ongoing cause, explained that he plans to bring a lawsuit against NGG, its board members, and Australia. The government of Australia, he complained, did not close down but profited greatly from Wau's gold. Like others, he viewed the early gold rush and long-term mining without acknowledgment of the valley's first residents by NGG as unfinished. The wealth created by gold still circulates. It is apparent in the structures left behind in Wau, and it is apparent overseas.

A favorite story about this wealth envisions it going into building the Sydney Harbour Bridge. The origin of this story eluded me during my research, but the bridge's grand scale certainly reveals differentials in the relations between Australia and its colonies. Furthermore, construction began in 1924 and was completed in 1932, overlapping with the early Morobe gold rushes and the development of NGG and Bulolo Gold Dredging.[5] Some see this wealth as part of a long-term exchange, a gift of sorts but with a greedy exchange partner. Going to court is the only way to seek redress, refolding events of the past into the present. In light of this ongoing attempt to seek compensation, can we really say that NGG has successfully closed its mine? Or rather, given their appearances, can frontier mines (and other conservation or development projects) ever fully close?

SCRATCHING THE GROUND FOR GOLD: TIME AND SPACE IN MINERAL EXTRACTION

One sunny evening in 2005, Yalamu was animated, acting out his words for our entertainment. His arms were folded and flapping at his side, feet scratching the sawn timber flooring as his head bobbed up and down. "Us landowners become like chickens, scratching and scratching," he proclaimed. We all laughed, and he repeated his phrase before adding, "Mining makes the ground useless." In contrast to timber, gold, he explained, does not regrow, and the land can only be a mine. We had been discussing New Guinea Goldfields and the small-scale artisanal mining that continues in its wake.[6] Indigenous alluvial miners continue to rework these streambeds, with entire families visible from early morning, waist deep in the silt and sand of past mining efforts (Moretti 2006). But more than an evaluation of artisanal labor, it was his analogy for what neighboring Winima villagers will be doing after the newly developed mine at Hidden Valley closes: they would make a living in much the same way.

Winima, Kwembu, and Nauti had signed the memorandum of agreement in support of Hidden Valley Gold Mine earlier in the month (August 5, 2005). The mine, located at the head of the Watut and Bulolo Rivers, was expected to produce approximately 285,000 ounces of gold per annum. However, following a temporary closure in 2017, technical difficulties, drought, fatal accidents, and a shifting partnership have reduced these estimates to 180,000 ounces (Harmony 2017). Though small in scale when compared to Ok Tedi, Porgera, or Lihir, development of the mine represented continuity in the exploitation of minerals along the Upper Bulolo. Like others in Elauru, Yalamu did not expect to see any substantial compensation or benefits from Hidden Valley. He was also critical of the present agreement and expected Winima to squander an opportunity for even greater gains. In the end, they too would become like chickens scratching the ground in search of gold.

Ours was a familiar scene during my time in Elauru, as we stood on the veranda of Mr. Waipi's house with a group of lively folks. Waipi, who had passed away many years before my introduction to his family, had built his house during the final days of New Guinea Goldfields as it diversified into logging along the Upper Bulolo and its tributaries. His house was one of those removed from NGG's logging camps and rebuilt in Elauru. That Yalamu was scratching away at the thirty-year-old floors complaining about past mining captures as much about temporalities and changes in development practices in Papua New Guinea as it does about mines and their closings. The house retains the urban appeal of NGG buildings, marking mining's potential as well as its past. Yalamu's complaint also raises an important question: if Biangai are experienced in dealing with mining, then why are they not better preparing themselves for the eventual closure of this most recent mine?

Since production began, a number of employees have built permanent houses in the village and urban settlements, but many rent homes at high cost in Wau, Bulolo, and Lae. Until recently, industry experience in planning for closure has largely focused on environmental and internal corporate financial planning (Chocilco 2002), with little attention to the impact that closure might have on communities in the developing world (but see Jackson 2002). While communities are increasingly included in contemporary closure planning, the general assumption is that mining ends when "the company" leaves. Along the Bulolo River and its tributaries, especially at Koranga and Edie Creeks, this has never been the case. Companies leave the area, but mining continues. For example, in 1981, NGG sold a controlling interest in their mining leases to Renison Goldfields Consolidated, which eventually relinquished these to Edie Creek Mining Company, a joint venture between the Biangai Development Corporation, Kukukuku Development Corporation, and the expatriate-managed Melanesian Resources. Mining continues there to this day under control of Niaminco (Australia). One could reasonably speculate that Hidden Valley will be no different. Biangai expect as much.

Mine closures do suggest dramatic changes in political, legal, social, and economic relations defined through places. For many, the departure of the company means an endgame to the extraction process—a disarticulation in a world of supposedly increasingly articulated parts (Ferguson 1999). Closure is a political and legal arrangement, planning for the end of corporate and even state obligations to the land and communities. It is an economic arrangement, altering the financial opportunities and fortunes of those once involved in resource exploitation. For employees of the company, investors, and others who benefit from its revenues, this is certainly the case. Thus, it is also a social change, severing relations among skilled and unskilled workers, middle-class employees, executives, and financial backers. Mine employees utilize their experience to transfer to other mines, continuing their relationship to mined places in a new locality. For these relations, closure planning certainly plays an important role in minimizing impact. But closure does not finalize relations between the resident populations and mined places, nor does it end the desires instilled during production. While the company might end its relationship to place, Biangai are forever folded into this transformed landscape, where their very personhood is linked to myth, history, and more-than-human socialities. Recovered, replanted, and refilled, mines remain important sites for imaginative engagement, for contestation, and for multiple understandings of their creation and dissolution long after the company has left.

Closure planning is about continuity and contestations over what continuity means. Mineral companies and governments view the impact of closure on communities and the environment in terms of postmining rehabilitation and sustainability (Chocilco 2002). For them, continuity captures the attempt to regain some of the pre-mine social and environmental conditions while carrying forward some economic benefits from the operational stage. Academic positions have largely reinforced the centrality of the company to closure planning, though with calls for greater attention to community and regional impacts.[7] This attention to companies is warranted to increase corporate attention to the impact of their actions. To this end, closures are supposed to be controlled activities, but they are written and implemented by underfunded staff and often with little government support (Burton 1995). But increasingly, closure is recognized for its affective (Pini, Mayes, and McDonald 2010) and long-term environmental impacts (Kirsch 2006, 2014).

In contrast to efforts to re-create, restore, and revitalize mined spaces, closure is also about the company's efforts to ensure its own continuity. Closure is needed to obtain a continual return on investment. As Harvey notes, increasing the speed of capital increases both production and profit (1982, 86). Mines that do not close and continue to work less feasible reserves slow down the speed of circulation and decrease rates of profit. Closures allow for corporate renewal, as companies move on to new frontiers. NGG ultimately succumbed to such pressures.

For communities, continuity can also mean continual mining of depleted resources, using small-scale alluvial methods or seeking out new corporate entities and technologies or even exploiting new areas to continue mineral exploitation (Cochrane 2017). It is about belonging to place and the attractive hold of past activities. Continuity here is not an attempt to mediate the socioeconomic conditions of mining and subsistence, but rather a desire for mining in perpetuity. While industry experts might view such continuity as "interference" in their planned demonstration of corporate responsibility (Chocilico 2002, 9n18), it also reflects different attitudes toward mineral resources and postmining development options. What is striking about the cases discussed here is not the impact that "closure" itself has had on Biangai, but the way the image of full production before closure—its appearance—has left a lasting impression. Theirs is not an imagined cargo, but it was very much a part of lived experience. Communities rightfully wonder why they must take a step backward while the company and its investors move forward.

This is equally true of conservation. After the community ended its relationship with Wau Ecology Institute, community members disassembled the buildings at the field station and divided the materials. The Kuper Range Wildlife Management Area was officially closed when community relationships through conservation places became divisive, but discussions of conservation persist. The practices of guiding "tourists" remain evocative, and during each research trip, former guides discussed with me their hope of restarting some form of conservation effort. The forest remains, they explained, and tourists still come to Wau. One or two a year find their way to the headwaters and secure access to the forests around the Kuper Range through the same guides that once worked with Wau Ecology Institute. Many come to hike the Black Cat Trail used by the Australian military during WWII. Biologists and other scientists still find their way into the area, attracted by its long history of research, if not by the presence of institutional support. Conservation places have not lost their appeal, and closure seems less than permanent.

Biangai attitudes toward closure and continuity also speak to different perspectives on temporality. Like other Papua New Guineans, Biangai contrast "their time" (*taim bilong mipela*) with "white time" (*taim bilong ol waitman*). In the Biangai language, *teng* is a recent construct to speak of the time of day. *Yok*, the sun, was the only marker of daily temporal change in the past. While *teng* has become necessary for school, church, and urban employment, *yok* speaks to the needs of subsistence gardening.

These contrasts take on a moral dimension, as punctuality, efficiency, and time management become central in ideas of postcolonial development (Smith 1994). Mineral extraction operates under such morality. Biangai employees have learned to distinguish between their work time and that of the company, and some aspire to mediate these realms (see chapter 5). However, while Biangai willingly distinguish the work time of the company from that of subsistence, they also emphasize the extension of time across space (e.g., Māhina 2010). As Wassmann (2001, 50) notes for the Nyaura (Middle Sepik), the geographical placement of events is of greater importance than their temporality. While Biangai historical events are sequentially ordered, where they appear on the ground is as important as where they appear in time. Kawna, with whom I began this book, reads this landscape with events in mind, demonstrating the power and efficacy of "place-thought" (Watts 2013). As a result, Biangai space and time are folded or gathered, places of myth and mining intertwined with gardened spaces and hunted paths. "Their time" and "their space" percolate with these pasts (Serres and Latour 1995).

Biangai mytho-historical time envisions a sequential understanding of how things and peoples came into being, which intimately connects them with their land and identity. The movements of ancestors, historical places, and sacred sites are spatially mapped as a flow of *appearances* across the contours of the land. They attract gardeners to their spaces through the intimacies of genealogy and knowledge of the landscape. Likewise, they repel those who cannot make such connections. I have argued here that Biangai also see place as attracting non-Biangai for logging, mining, and conservation. Place is not simply a location of connection but an intimate part of what it means to be a person and to reproduce society.

Biangai time is also cyclical, following the annual planting and harvesting of subsistence crops and wild foods, the seasonal movements of animals and changes in weather.[8] Here events, practices, and things do not finish, but repeat daily and annually, accumulating onto the intimacies of the past. As Nancy Munn (1986, 9) has effectively argued for Gawan exchange networks, the acts themselves, as spaciotemporal practices, entangle both persons and places. Appearances take up the congealed form of such actions as an aesthetic presence that is greater than the object exchanged, fallow garden, or other mnemonics.

For example, annual cycles of planting reinforce and renew genealogical, mythical, and spatial relations. Planted by men and tended by men and women, yams are likened to children. They too have genealogies, propagated from past seasons' crops and folded into gardens that are still marked with the small but noticeable "sinks" of the last time the land was tilled. Their later appearance as secondary forest spaces likewise *attracts* gardeners, bringing together the families of those who once planted there.

Each kin group is likewise connected to the land through mythic topologies, highlighting ways in which mytho-historical and cyclical temporalities mutually inform one another. Throughout Papua New Guinea, these mythic pasts are remade into the present through ritual.[9] Such rituals occur at sacred sites variously referred to as "the spring" (Umeda; Gell 1975), "the knot of the land" (Huli; Goldman 1983), or the "earth joint" (Paiela; Biersack 1999), evoking metaphors of growth or a folding of the land. Like the practices associated with *ruwera*, time and space are implicated as distinct points of ritual significance. At such spatiotemporalities, past, present, and future are gathered. They are "point[s] of calamity" and "point[s] of conservation, where fertility rituals were performed" (Biersack 1999, 73), where human-environment relations are renewed. While these sites of mythic power continue to appear in contemporary practices (Jacka 2001), such folding of the past into the present is not just grounded in spatiotemporal rituals. It is also featured in more quotidian practices, such as hunting, fishing, and gardening.[10] Throughout, it highlights a conceptualization of time as gathered into the earth's contours. But more importantly, it is the folding of social relationships into such spatiotemporalities that makes place meaningful.

While Biangai have forgone formal rites of renewal, the significance of renewal for daily life continues. In negotiating a proper cognatic marriage, Biangai consider what land each partner can contribute and where it is located. Spatial proximity of resource rights makes for a good relationship, "gathering the ground" (in Tok Pisin, *bungim graun*) of past activities by the ancestors of both marriage partners. The exact time of use is secondary to location. Marriage enacts these relations, reuniting parcels of land that are in turn used in an annual cycle of planting and harvesting. The practice "reveals a time that is gathered together, with multiple pleats" (Serres and Latour 1995, 60). It is the parcel's appearance as a viable garden site for a couple that attracts human social relations. Children born of such more-than-human relationships with place speak of their *ngaibilak*, placing them in a caring relationship with the land. Marriage is as much about the aesthetics of place—its appearances—as it is about the aesthetics of persons. Planting a garden becomes an act of mutual incorporation, where subsequent plantings evoke these pasts, remaking their surfaces into present practices. The land is replete with the past, giving it a powerful and persuasive appearance in the present. What is of interest is the role of these open and cyclical temporalities in Biangai experiences of mineral extraction.

The environmental destruction left in the wake of Ok Tedi's failed tailings dam required Yonggom to envision "new forms of chronology . . . that transcend place" (Kirsch 2004, 202). Time replaced space as space was devastated by the waste that continued to flow down the Fly River. It is not the appearance of successful mining that concerns the Yonggom, but its repulsive qualities. For Biangai, however, mining has not erased their perspectives on spatiotemporal relations, and it continues to be attractive. In either case, the aesthetics of the past inform how the present is experienced. Such practices continue to influence how Biangai anticipate the potential for present mining, both real and imagined.

LOOKING TO THE FUTURE: SCRATCHING THE EARTH AGAIN

It is almost a cliché to say that companies and not the government in Papua New Guinea bring development and basic services to communities (see Golub 2014; West 2006). This adage is certainly reflected in Wau's mining history. For Biangai, early gold rushes produced little income. Changes in the relations of production wrought by independence have significantly increased their returns compared to previous generations. This is not because of higher-valued reserves—the reserves are increasingly marginal and require greater financial and technological investment to extract—but because they have fared better and negotiated better, and they now have government-established rights. A closed mine, or even the threat of closure, redirects community energies toward finding a better mining partner, if not toward the court. Postmining and postconservation spaces still appear to be full of potential in such landscapes.

An Elauru man who joined us for Yalamu's chicken impersonation complained that Winima would fail to get the most out of their current mining partners at Hidden Valley. At first I misunderstood his point, and he explained that the compensation from gold must be split four ways. Winima and Kwembu would receive one share, while the government, the company, and the Watut communities would divide the rest. He argued that if all Biangai communities were fully included, Winima and Kwembu would increase their return as well. It was a calculus that would make little sense to frontier capitalists. But the crux of his argument, and the argument in many of the communities outside the memorandum of agreement, is that working together strengthens the Biangai position; while humans are attracted to gold, the gold only *gives* itself when all those with connections to places are acknowledged. There is a certain truth in this interpretation, as mining hinges on regional stability, and dissatisfied community members have frequently disrupted mines (Filer 1998). However, as with gardens, marriage, and other grounded practices, my friend wished to emphasize the gathering of the land. He wanted to recognize that all Biangai are related through the land of the mine. Furthermore, in one particular folding of the past, compensation would be more widely distributed and the ore itself would be revealed more fully. Either way, he argued that there would be further mining in other locations along the folds of the same ore body.

For Biangai, the courts have provided one solution to these debates in the past, as they contest the degree to which Hidden Valley's location connects to all Biangai speakers. As in Biangai efforts to win compensation from NGG, courts allow for the expression of ancestral ties, renewing relations that have become less "active" in recent times. Movement, internal migration, and intra-village marriages mean that many can speak of this area, relating it to the places and paths (*kasi mek*) used by their ancestors. Time has not eroded these connections, and courts are seen as an opportunity to express these spatiotemporal relations.

In addition to considering legal action against Winima and Kwembu for rights to Hidden Valley, the other five Biangai villages either have or desire their own projects (consider also Jorgensen 2006; West 2006). Development is invested in the land; this has already been demonstrated to them. Now it is just a matter of reconfiguring the relations that make it possible again. In 2005, a senior Elauru man with strong connections to Winima complained about the distribution of benefits to other Biangai. When I asked if he would want to see a mine open in Elauru, he explained that they would negotiate a better compensation package because of their strong leadership. Perhaps this will not occur during his lifetime, but eventually. Likewise, the neighboring village of Werewere was working with a company they called Terra Nulle to test the nearby Lake Trist area for nickel. Leaders from this village insisted that they too would develop a better relationship with the company. The company, whose name that means "no-man's-land," was promising everything the community desired. During the initial meeting, where they welcomed the company's management in "traditional style," the director commented to them that this was the first time they had received such a greeting. For the company director (according to my friends in Werewere), it demonstrated Werewere's commitment to place and the project. Werewere

Biangai felt assured that this promised good social relations with the company. The project never eventuated, but as recently as 2016, Werewere residents were still discussing the potential.

Biangai know that a given resource is finite. They know that mining companies will eventually extract the bulk of the reserves and, failing any new finds, close down the operation. But they also believe that other reserves will be identified in other places, and they will be able to renew relations with a new company. The appearances of past mines are too powerful and too attractive to forget, and potential mines are envisioned across the landscape. In 2001, the mood in Winima village reflected this sort of optimism. In interviews about the postmining community, stakeholders and non-stakeholders alike foresaw the urbanization of village life (see also Jacka 2001; Kirsch 2004). Power lines, permanent homes, ease of access to stores and commodities, transportation, government services, medical facilities, and school opportunities would be readily available. Those who did not see the transformation of village spaces imagined formal urban migration to Wau, Bulolo, and Lae. These images were still pervasive as Philip imagined his future house. Like NGG's employees and shareholders, Biangai would eventually achieve the sort of modernity that Wau once represented.

In comparison to previous mining ventures, Winima had been quite successful in gaining access to the companies that prospected the Hidden Valley site, placing the community into the calculus of what makes mining possible. This strategy fueled development hopes, refolding the appearances of past mining into their present engagement with companies that have worked at Hidden Valley. For my Winima friends, closure would not mean an end to company benefits but would lead to the same opportunities that colonial mining had made possible for its investors. These were imaginary futures, made visible through their reading of Wau's past, NGG's longevity, and the continuity that mining brought to a town that grew out of the forests. Given Wau's recent decline, only gold could provide services once again.

The reality of diminished reserves relative to the pre-WWII gold rushes did not set in until 2005, after the memorandum of agreement was signed. Winima villagers had recalculated their futures, assessing the potential revenues and compensation to be substantially less. In 2005, the mood was one of scratching at the ground and trying to make do with what little would be left. By 2015, Philip, my host father in Winima, explained that earlier they thought they would all be wealthy, that things would change a lot as a result of mining. As he complained about his house and the ongoing violence between families in the community, he lamented that for now "we are going to stay the same," gesturing toward the rest of the village. But he felt that prospects were good for mining on their land nearby at Kudjuru.

• • •

Wau's story could be told about many mining communities in Papua New Guinea and many widely scattered development projects worldwide. Throughout the world, mining is implicated in environmental destruction (e.g., Harper 2005; Kirsch 2006, 2014), civil war (e.g., Nash and Ogan 1990), labor exploitation, and marginalization (e.g., Jorgensen 2006; Ferguson 1999), dividing communities and disrupting relations among kin (see Banks and Ballard [2003] and Godoy [1985] for reviews of these concerns). And yet, mining remains attractive in a place like the Upper Bulolo. While compensation figures into the aesthetic appeal of mineral exploration, this too is often illusory, as benefits are unequally distributed or quickly squandered. Like postsocialist nostalgia in Eastern Europe (Berdahl 2001) or longing for a return to colonialism in Africa and elsewhere (Bissell 2005), the appeal of mineral exploitation is more than simply about the lasting impression of a set of symbols over time. It is about the sociocultural *and* physical transformations of the past into an imagined future. There is a bit of nostalgia in Biangai longings for continual gold mining, but in contrast to the "longing for a return to something that cannot be restored" (Bissell 2005, 225), mine closures often lack permanence.

The sorts of transformations in community desires that occur during mineral exploration and exploitation are not easily forgotten in a postmining, postdevelopment environment. The appearances of past mining remain close at hand, folded into the contours of the land. It is these appearances, attractive or ugly, that continue to draw them into the present. Closure does not end such desires, but the expression of desire differs for each stakeholder. The company fulfills them by moving to new resource frontiers, performing closure as a rite of corporate renewal. The communities that are left behind cannot make such a change. Instead they look for new opportunities to acquire the appearances of past successes—to fold these moments into present practice to realize the potentials of place and their more-than-human world. In many instances, small-scale artisanal miners eke out a living panning the streams and scratching away among the wastes of previous mines. But experience in Wau and elsewhere also suggests that mines reopen as technologies shift or new companies, new practices, and increased commodity prices allow new economically feasible reserves to be identified. Mining continues until the entire ore body is ripped from the earth.

For Biangai, closure is not so much about the end of mining and the reclamation of the places for other productive endeavors. Instead, closure is about a change in the social relations that make mining possible and the promise that the appearance of a new mine brings. These kinds of relations with places inform community ideas about development efforts and have a long-term, seemingly endless impact on the local stakeholders. Companies and NGOs model development, inscribing the local landscape and practices in ways that are difficult to erase. But this is not through the permanence of the physical transformations. Instead, the practices of development are made even more permanent through their very appearance within local spatio-temporalities and the continual attempt to gather the past into the present.

Experiences with mine closures have taught Biangai that there is compensation and continuation—that mines reopen, that while the bankable reserves might be extracted, the resources remain. Closure is an opportunity for negotiation, for the ceremony of the courts, the exploitation of less feasible reserves, and the identification of new sites for exploitation. Closure is gathered into the folds of localized timescapes, where appearances of past activities remain. It was evident in Winima's anticipation of Hidden Valley and their disappointment when more realistic returns were presented. For both Elauru and Winima, the future is about the ore body extending far beneath the surface. Its full character has yet to be revealed, and eventually, because they are the right ones to establish relations with the gold, they await the next mine.

CHAPTER SEVEN

Belonging

NEAR the end of my research visit in 2016, Mama Sabu asked a seemingly odd question: "Where are you from?" Since I had spent thirty months in the community, I first thought that she must have forgotten the name of the place. As I looked around the room, the firelit faces of the family looked at me with genuine interest as I said, "I'm from Kentucky, I guess." I explained that while I've lived in Minnesota, Papua New Guinea, New Zealand, and Texas, I still consider myself a Kentuckian. I was born there, I explained. Sabu thought for a moment, then followed up her question with the real point of her query: "But where are you *really* from? Where does your family *belong*?" Belonging to a place is something more than simply being born there. Sabu and her family belong to Winima and Elauru. It is where they have inalienable connections to gardens, to trees, to places, and to each other. These are the things "that sustain one's life" (Wood and Winduo 2006, 85) and define Biangai "identity, belonging, and social existence" (Stella 2007, 29). My own sense of being from Kentucky was not the same. My ancestors came to Kentucky between three and four generations ago, tracing a path from all corners of Europe. We didn't have permanent inalienable rights to land and resources. We bought, sold, and rented places to live in as we needed them. Nothing was permanent. These places—lands made meaningful by Shawnee, Dakota Sioux, Maori, and Coahuiltecan peoples—were problematically entangled in a history of colonialism and conquest.[1] My own ancestry traces through both Western and Eastern Europe, among peoples whose own lands were enclosed by monarchies and authoritarian states. While Biangai view themselves as emplaced, my family history was most certainly one of displacement.

Philip thought about the conversation and astutely observed, "This is what is happening to us." As Biangai marry into other villages, other language groups, and even other nationalities, Philip observed that their children are *hapcas* (half-caste), belonging to two places. Biangai often see this as good, creating paths (*kasi mek*) between locations, much like the paths created in the past between Biangai gardens and villages. But if the children and grandchildren marry in the same way, adding new places to the genealogy, where do the future generations really belong? Everyone was silent as the idea sank in. Belonging to place was not certain for future generations of Biangai, Philip feared. The factors driving these movements are not that different from those of a fourth-generation Kentuckian of mixed European ancestry. Economic mobility, job opportunities, marriage, family, disputes, and cost of living figured prominently in recent urban migrations. The real dangers to social reproduction of the community are for the next generation of urban Biangai, with commitments to other places and other peoples. For those who remain *placed* in the social life of the community, hope resides in the land. But for those who move away, hope is displaced from such commitments and tied to work, education, and the seemingly distant aspiration of finding success in business.

Place is important in Oceanic world-making. Places are not simply locations where things happen, and places are not made possible only through human social relations. Places and persons are entangled in a mutual becoming, making it hard to separate them conceptually. For Biangai, relations with and through place are what make marriages possible, foods grow, minerals emerge out of the ground, and researchers come into their lives. Furthermore, these relations shape the contours of successful development. If successful, coffee would carry the Biangai name as the source of the best organic, shade-grown coffee in the world. It would bring businesses and high prices into the Bulolo Valley. If successful, conservation would link global scientists and tourists to Biangai places of importance. The dead insect bodies that circulate in gift shops and museum collections would be meaningfully tied to habitats, entangling them with valuable relationships. Companies, such as Givaudan, would properly recognize Biangai knowledge of the scents they cultivate. If successful, mining would transform social life, making it possible for Biangai to move easily between places in Papua New Guinea and abroad, while transforming the village into a modern Pacific community. Royalties and employment would enrich their connections to place rather than sever them. However, as Sabu and Philip fear, the boundaries between place and person become difficult in a world in motion. While yams, coffee, family, home, and a sense of belonging continue to bring people back to the community, will they always?

NEOLIBERAL PLACES

Questions about our relationship to place are extremely important for people throughout the world. Many have noted the pressure on peoples to transform their relationship to places and each other as a result of resource extraction, changing labor markets, and capital's need to find new opportunities for profit. Neoliberal capital requires the "creative destruction" of "prior institutional frameworks," including "divisions of labor, social relations, welfare provisions, technological mixes, ways of life, attachments to the land, [and] habits of the heart" (Harvey 2005, 3). These practices have set people in motion. Places and people are detached through economic, ecological, political, and social upheavals. But likewise, products of the land circulate with tenuous ties to their places of production (e.g., West 2012), and while we might value biological diversity, valuing places of diversity continues to be difficult to sustain. It is not just a revaluing of "nature" that should concern us but the revaluing of ourselves in relation to the world. Biangai are confronting the possibility of such dislocations of placepersons, while at the same time suggesting that the "destruction" does not have to be absolute.

These are not naive perspectives made without experience in the extractive economies of mining and the less innocuous practices of conservation. They are made in the context of an ongoing connection to place. This is not simply an attachment but a mutuality that has yet to be severed (Harvey 2005). They don't have to "scratch the ground" for benefits. According to Biangai, the benefits from the land are rightfully part of their very being, whether in the form of yams, coffee, flora, fauna, or gold. Our connections to place are thus important for understanding both our relations to each other and our relationships with the plants and animals with whom we share the world.

What Biangai experiences suggest is the power of place as a central condition for their ongoing sense of hope. Extending Eben Kirksey's rgument that "emergent ecologies" are "flourishing in the aftermath of order-destroying disruptions" (2015, 217), Biangai experiences suggest that such an ecological emergence is made possible through their relationships to place. Places are mined for resources, observed for their flora and fauna, engaged because of their diverse human populations, but they are equally entangled in our very becoming as the source of food, histories, relationships, and life.

In spite of Biangai optimism, I can't help but see capital's "iron cage" (Weber 1930). There are powerful institutional forces that press for Papua New Guinean places to be privatized, or at least registered in such a way that

they are more accessible to capital and perhaps more accessible t
This is an old debate (e.g., Marx 1979; Wordie 1983). Postcolon
establish lease relationships, associated with logging, mining
expansion, are certainly part of this process of enclosing land. Ongoing policy discussions in Australia about the role of aid to its former colony have emphasized private property and corporate registration regimes as positive interventions in Pacific Island communities (Hughes 2004). Echoing the "tragedy of the commons" thesis in spirit, if not verbatim, neoliberal policy "experts" advocate for privatization as a "necessary precondition if the people of Papua New Guinea are to aspire to the higher standard of living that only becomes feasible under private landownership" (Lea and Curtin 2011, 1).

After foreign policy advisers and national leaders in Papua New Guinea failed to promote a formal registration of land, government reforms emphasized the registration of landholding groups. In either case, security of investment is the central concern. Recent attempts to formalize landholding groups have combined the voluntary organization of local communities with the creation of a secure lease agreement that is made transparent through a centralized governing body (Chand 2017). The result would still effectively allow the alienation of places from fully local engagements through lease agreements. As a result, place can become subject to rational calculations, monetized efficiencies, and external control, while the persons who are made through places are transformed into rational collectives, if not rational actors.

Biangai subjectivities are transformed in their encounter with both mineral extraction and conservation, as well as ongoing entanglements with coffee and other cash crops, but it is their connection to place that maintains their sense of hope and their connections with each other. As their debates around mining and conservation demonstrate, the denial of shared rights to a communal space is a source of contention and disruption. Monetization of relations might be the primary way to improve quality of life for neoliberal subjects, but this is not the standard of living that Biangai are looking for. They hope to maintain their connections to place, while improving the quality of *community* life. As my conversation with Sabu and Phillip suggests, a higher standard of living can be more than just rationalization and monetization of relationships.

Development of Biangai resources should not be only about benefiting corporate stakeholders, NGO staff, and others who lack a commitment to Biangai places. For placepersons, the extraction of resources and practices of conservation occur only because of place-based histories, intimacies, and connections. They turn on its head the idea that security of investment requires alienable land regimes. Instead, their inalienable connections to the physical world are what should make development possible, as real connections to place are ultimately sources of both sustainability and security. Recent ethnographies of mining in Papua New Guinea have highlighted the differences between Papua New Guinean and global mining practices and the differences between what communities expect and what they receive.[2] The expectation of benefits is not unreasonable, as capital makes demands on their lives and lands. "Whose development?" is an important question for us all, as it places stakeholders in relationships that are grounded in local and indigenous rights.

In spite of hope for the next opportunity, the strain ripping apart place and person is most apparent in these communities. Violence, divisions, fences, urban migration, commitments to employment outside of the community and its sociality, and growing economic inequality are all symptomatic of capital's creative destruction. Displacing Biangai and turning the soil against local spatiotemporalities, extractive industries promise just what their name implies—a separation. But for now, places still hold sway with the promise of the return and renewal of relationality.

Biangai world-making raises some important questions for the rest of us. What if we didn't see the world through capital's destructive lens? What if scientists visiting communities celebrated local knowledge as much as the presence of endemic species, acknowledging that both relationships are made possible through place? Can we imagine a world where drinking a cup of coffee (or tasting some other commodity) evoked a real intimacy with the places and persons of their production? Or where we sought connections through the precious metals that are folded into our technological networks, ornate jewelry, as well as mundane consumptions? Placing our own personhood in such a world is an intriguing possibility that Biangai and wider Oceanic imaginations suggest as a more hopeful way of relating to the world.

NOTES

INTRODUCTION

1. See also Refiti (2017) on Bergson's theory in relation to Oceanic concepts of time and space.
2. While Vayda has emphasized the significance of events as a focus for ecological anthropology, here I'm thinking about the potentialities of a would-be event. What might happen is just as important as what does happen.
3. Likewise, in discussion of an Oceanic theory of time and space (*tā-vā*), Samoan architect Albert Refiti (2017) notes that specialists such as the ritual expert or tattoo artists are able to make space mediate the rhythm of time in Samoa.
4. See also Gordillo 2004, 2014; Tsing 2000.
5. The idea to combine the two concepts came from conversations at Columbia University's Ecology and Culture Seminar. I would like to thank Paige West for the suggestion, as it very much reflected what I think Biangai are expressing.
6. E.g., Descola 1994; M. Strathern 1980; R. Wagner 1975.
7. E.g., L. Smith 1999; TallBear 2014; Todd 2016; and Watts 2013.
8. E.g., Gegeo 2001; Ka'ili 2017; Māhina 2010; Nabobo-Baba 2006; Roberts et al. 2004; Teaiwa 2015; Tuwere 2002.
9. See also Clark 1993; Jacka 2015; Wardlow 2006.
10. For indigenous North America, Watts refers to this way of knowing as place-thought, where people and the land co-constitute one another—tied not to our ability to grant agency, but recognizing that all things are "tied to spirit, and spirit exists in all things" and therefore "all things possess agency" (Watts 2013, 30).
11. This is demonstrated most recently in the legal status granted to Mount Taranaki and Whangani. For Maori communities, making these entities legal persons recognizes their status *as* ancestors.

12 Büscher and Davidov 2013; Doane 2012; Igoe 2017; Fletcher 2014; and especially West 2006 have developed this connection in some detail.
13 Castree 2010; Hilgers 2011; Peck and Tickell 2002.
14 See Ferguson 1999; Harvey 2005; Kosek 2006; T. Li 2007, 2014; Raffles 2002; Sawyer 2004; and Welker 2014.
15 For the range of positions taken, compare Boserup 1965; Clarke 1976; Conklin 1957; Geertz 1963; Netting 1993; Polanyi 1944; and Taussig 1980.
16 E.g., Agarwal 2005; Benediktsson 2002; Dudley 2002; Errington and Gewertz 2004; Gupta 1998; Hann 2003; Mackintosh 1989; Scott 1999; West 2012; and Wittman 2011.
17 E.g., Godelier and Garanger 1979; see Netting 1974 for a review.
18 But see Banks 1999; Godoy 1990; Hyndman 1994; Ogan 1996.
19 E.g., Garvin et al. 2009; Himley 2012; Welker 2009, 2014.
20 Mining development variably offsets agricultural work with employment and other forms of compensation, suddenly creating a new class of workers, new inequalities, and new subjectivities (Imbun 2000; Nash 1979; Taussig 1980; see also reviews by Banks and Ballard 2003; Jacka 2018). While we know that extractive industries also dramatically alter food pathways or provisioning (Narotzky 2004), the processes have remained underinvestigated. This has particular significance for future food security, as gold mines do eventually exhaust their reserves and community incomes dramatically decline with closure. Historically, there has been a great deal of attention to changes in the relationship between households and land use (e.g., Chayanov 1966; Netting 1993). But, as Perz (2001) notes, farming communities are not homogenous, and multiple factors must be considered in different cultural and ecological contexts. Here attention will focus on shifting household dynamics as mine workers, ecotourist guides, and their families try to maintain subsistence and cash crop agriculture. At the regional level, extensive research has been conducted on trends in agricultural production and consumption. Agricultural specialists, geographers, and anthropologists provide a good overview of macro-level practices (e.g., Allen et al. 1995; Bourke and Harwood 2009; Gibson 2001) and attend to the transformations of a purely subsistence-based national economy into one that favors smallholder cash crops (Allen et al. 2009, 293; see also Benediktsson 2002; Errington and Gewertz 2004; West 2012). However, research that examines local motivations and practices in agriculture transformations in Papua New Guinea are few (but see Nordhagen et al. 2017), especially as oil, gas, and mineral extractions account for 75 percent of GDP.
21 See Blackwood 1978; Burton 2000b; and Moretti 2006.
22 Watut is a name given to a diverse group of Hamatai speakers (within the broader Anga linguistic family). During the early gold rushes along Edie Creek, miners and patrol officers referred to them as Kukukuku, a name

that remains prominent in the ethnographic literature, as Blackwood conducted fieldwork among Upper Watut communities between 1936 and 1937 (see Blackwood 1978).

CHAPTER ONE: MINING NATURE

1 The name Guinea is probably derived from the name of the trading town of Jenne near the mouth of the Niger River along Africa's west coast. As a result of the gold found in the area, Guinea became a common name for gold coins (Bernstein 2000), and Nuevo Guinea appropriately marked the potential that explorers saw in the islands.
2 For further discussion of this debate, see Arens 1979; Creed and Hoorn 2001; Dixon 2001; and Obeseyekere 1992.
3 Oldörp and Dammköhler are said to have discovered gold along the nearby Watut River in September 1909. They were reportedly attacked by Middle Watut peoples, leaving Dammköhler dead and Oldörp wounded (Healy 1967).
4 See Burton 1996b; Martin 1992; Ouellette 1987.
5 A bend in the current road between Wau and Bulolo is still referred to as Missus Booth's.
6 Samuel James Appleby, "Report on the Native Disturbances, Biololo (Rabaul) Area," NAA 1927/728, National Archives of Australia.
7 Willis (1977a, 1977b) summarizes the history of the Kaisenik Killings.
8 In a report submitted by Assistant District Officer Appleby, he reported that the first two bodies were buried by one of the mission helpers. No official report of cannibalism exists for the other three.
9 Of Idriess, J. Sinclair notes, "When I was living in Wau in 1948, old hands assured me that they had spun Jack Idriess a lot of colorful yarns, many of them true, others not" (1998, 30). Charles Wells, chairman of New Guinea Airways, likewise expressed doubts about the quality of both Clune's and Idriess's works, noting in a letter to the editor of *Pacific Islands Monthly*, "What fills me with depression is that crude writers of the type of Idriess and Clune should be the only ones game enough to tackle this colossal story" (Wells 1939).
10 See Willis (1977a) for further discussion. In 2001, I witnessed a similar display by elder women following the death of a village councilor. They believed that he was killed by sorcerers and provoked men in the village to take action against the accused.
11 Appleby, "Report on the Native Disturbances."
12 See discussion in chapters 2 and 3; see also Fortune 1963; Harrison 1982; Tuzin 1972 for similar personifications.
13 He was correct, as the company (Big Bean Coffee) lasted for less than ten years before being forced to close. As of this writing, the plantation

remains unused, overgrown with vines, and picked by random urban residents to make some extra cash. The town council still hopes to develop it.

CHAPTER TWO: GROUNDING KINSHIP

1 While Biangai elders wanted me to know their story of creation, they also wanted the details kept secret. I was told that I should not mention specific names and places, which are even more significant as these remain important as testaments of customary rights and contentious issues of compensation.

2 Biangai emphasize that though their ancestors could appear as animals (and still do), they are fully human. This distinction is important in how Biangai separate themselves from other peoples (i.e., Watut, Buang, coastal populations), whose histories, they claim, portray the first humans as originating from animals.

3 Contrasted with Taussig's (1980) animatistic depiction of Mount Kaata and the totemic image of the Mount Kare python in the Southern Highlands of Papua New Guinea (Biersack 1999), Biangai conceptualize a more personalized spiritual realm. Biangai mountainscapes are differentiated with named ancestors who hold specific relationships with the living and the land. In an older typology, theirs might be called animism (see Bird-David 1999 for a review).

4 For an early discussion of this, see Fortune 1963; Mead 1935; Tuzin 1972.

5 E.g., Brightman 1993; Ingold 2002; Thornton 2015.

6 E.g., Carrier and Carrier 1991; M. Strathern 1990; R. Wagner 1975; A. Weiner 1976.

7 Consider also Carsten 2000, 2004; Haraway 1997; Roberts et al. 2004.

8 E.g., A. Anderson 2011; Hess 2009; Leach 2003; and Taylor 2008.

9 See A. Strathern 1971; M. Strathern 1990; Biersack 1999; Jacka 2015.

10 Nekgini husbands and wives "work" the land so that their children may grow from the abundance it produces. However, children lack "substantial connections to maternal and paternal groups" (Leach 2003, 149), meaning that they are not the product of distinct lineage contributions. Instead, as maternal and paternal relatives engage in exchange around the growth and development of a child, they are not compensating or substituting for the place of that child in their lineage; they merely make the relations visible. Nekgini easily adopt non-kin into their family as work and food define social relatedness.

11 Hence, persons and concepts of personhood change over the course of one's life (Udvardy 1995). However, places are meaningful genealogically *and* significant for reconstructions of historical events. Time is spatialized (Rosaldo 1980), and space is socialized. Such conceptualizations

have great implications for development projects that utilize and transform both the land and the labor of parents, requiring new regimes of work (see M. Smith 1994).

12 E.g., A. Strathern 1971; M. Strathern 1990; R. Wagner 1967; J. Weiner 1988.

13 Given Burton's work with the principal landowners of the Hidden Valley mine, this seems to be a reasonable reflection of tensions within the landowning communities. Winima landowners were more likely to make patrilineal claims during discussions of garden lands, and with respect to the mining site, they were often adamant about the primacy of paternal rights (Halvaksz 2005). Elauru, by contrast, was more committed to a cognatic ideal.

14 Territory of New Guinea, *Patrol Report, Wau no. 34/47–48, January 11, by D. Maclean, Patrol Officer* (1948), National Archives, Port Moresby, Papua New Guinea.

15 Territory of New Guinea, *Patrol Report, Wau no. 5/1961–62. January 23–27, by P. G. Whitehead, Patrol Officer* (1962), National Archives, Port Moresby, Papua New Guinea.

CHAPTER THREE: WORKING THE LAND

1 In Biangai, gardens and work are linguistically similar. *Yawe mizi* means "to work" while *yawe ligi* references a garden. Gardens are therefore idealized as places of work.

2 See Bell, West, and Filer 2015; Errington and Gewertz 2004; Filer 1997, 1999; Golub 2014; Jacka 2015; and West 2012.

3 E.g., Finney 1973; Keesing 1992; Rodman 1987.

4 Bashkow (2006) makes a similar argument for the Orokaiva and their relationship with taro. While other foods are prevalent and problematic, taro is an essential element of Orokaiva identity and their rootedness to place. Like the Orokaiva, Biangai readily adopted rice, instant noodles, tinned meats, and other foods into the social life of the village. But as discussed in chapter 5, the dichotomy "whites versus us" elides the distinctiveness of a range of foodstuffs introduced into mining camps.

5 During 2011, 2014, 2015, and 2016, I used a Trimble GEOXT 2008, achieving sub-meter accuracy post-correction.

6 El Niño climatic cycles impact Papua New Guinea greatly via dramatic variations in rainfall, resulting in food scarcity and increased frost at higher altitudes. While the impact on the Bulolo Valley is notable, the government and the mining company provide timely assistance. One major impact had been the reverse migration from urban areas, where water became scarce, placing greater strain on local food supplies. Relative to other regions, the severity of these cyclical events has been muted (Bourke, Allen, and Lowe 2016).

7 During my initial survey between December 2000 and January 2001, as one might expect, household income was much greater in Winima, with the average household reporting an annual of 1,981.88 kina (total village income for 2000 = 95,130.02 kina). Elauru reported a much smaller household average of 568.56 kina per year (total village income for 2000 = 22,174.00 kina). These numbers do not include mining compensation, most of which was held in an account until disputes over rights to the mining area were resolved.

8 Territory of Papua New Guinea, *Patrol Report, Wau no 3/1955–56. November 29–December 17, by G. R. Grey, Patrol Officer* (1955), National Archives, Port Moresby, Papua New Guinea.

9 Territory of Papua New Guinea, *Patrol Report, Wau no. 5/1961–62. January 23–27, by P. G. Whitehead, Patrol Officer* (1962), National Archives, Port Moresby, Papua New Guinea.

10 This does not include the twenty or more households in Kudjuru that are part of the village of Winima in terms of district representation. See discussion in the introduction.

11 Territory of Papua New Guinea, *Patrol Report, Wau no. 7/1968–69. March 17–April 24, by K. G. T. Sandell* (1969), National Archives, Port Moresby, Papua New Guinea.

12 Funding allotments are administered by members of parliament, who are often seen as potentially corrupt.

CHAPTER FOUR: BECOMING CONSERVATIONISTS

1 Initially founded as a research station for the Bishop Museum (Honolulu, Hawai'i), with funding from both the Bishop Museum and the Smithsonian Institution (Washington, DC), WEI slowly acquired sixty hectares of property adjacent to the colonial mining town of Wau (Wau Ecology Institute 1976). The current boundaries of the institute and of the town of Wau both rest on lands that were originally alienated from Biangai speakers by the colonial government, which Biangai continue to claim. By 1975, the grounds included office space, residential housing, a library, a laboratory, an arboretum, experimental gardens, museum exhibits, and a small zoo. Researchers affiliated with both WEI and the Bishop Museum began publishing extensively on regional flora and fauna (Wau Ecology Institute 1974, 1976). Following Gressitt's death, Harry Sakulas, a Papua New Guinea national from the Sepik region, was trained to take over the operation, which he ran until its demise in 2007 (Halvaksz 2005).

2 E.g., Ellis 2002; Kirsch 1997; van Helden 2001; J. Wagner 2005; West 2006.

3 Under the Fauna (Protection and Control) Act (1974), local landowners and NGOs used wildlife management areas to manage natural resources

in Papua New Guinea and establish sustainable economic alternatives to resource extraction. Unlike protected areas, wildlife management areas enable communities to maintain customary tenure while regulating the use of fauna. Recognizing the depletion of resources due to population pressures, external demands (e.g., commercial hunting, trespassing, squatting), and the introduction of new technologies that increase exploitation (e.g., shotguns, dogs, chainsaws), wildlife management areas are intended to help communities better manage their resources and respond to changes in the local economy. With the assistance of Department of Environment Conservation (DEC) officers or NGO facilitators, communities elect representatives that determine and mark boundaries, agree on rules, and establish small business ventures that ideally are ecologically friendly. The completed proposal is submitted to the DEC for approval.

4 Regina v. Deslisle, [2003] B.C.C.A. 196 (Can.).

5 The habitat is still threatened by expansive logging and oil palm projects. Conservation efforts continue with a mix of local, national, and international support.

6 Harvey elaborates: "Capitalism is, from this last standpoint, beset by a central moral dilemma: Money supplants all other forms of imagery (religion, traditional religious authority, and the like) and puts in its place something that either has no distinctive image because it is colorless, odorless, and indifferent in relation to the social labor it is supposed to represent, or, if it projects any image at all, connotes dirt, filth, excrement, and prostitution. The effect is to create a moral vacuum at the heart of capitalist society—a colorless self-image of value that can have no purchase upon collective as opposed to individual social identity" (1996, 156).

7 Simmel long ago highlighted that "value is never a quality of things, but a judgment upon them which remains inherent in the subject" (1978, 63). Thus, different subjects will constitute the value of insects in unique ways, informed by the qualities transferred and created in their motions. The place of contact is as important for value as the subjects involved.

8 Human-animal relations have been the subject of much recent theoretical development, as their agency both in ecological systems and as part of large networks become the subject of social scientific investigation. As companion species (e.g., Haraway 2003, 2008), entertainment (e.g., Davis 1997), feral pests (Knight 2003), and portraits of human societies (e.g., Haraway 1989; Ritvo 1997), animals enter Western society and discourse and challenge our own classifications and social relations (Levi-Strauss 1963). There is a great deal of emphasis in discussions of human-animal interactions on the intimacy of such relations for indigenous peoples (e.g., Halvaksz and Young-Leslie 2008; Ingold 1980; Feit 1998; Sillitoe

2002; Young-Leslie 2007), as well as Western urbanites' resolve to reconnect with the species and places of an idealized and indigenized nature (e.g., Braun 2002; Harvey 1996). Insects remain minor contributors to this theoretical development. In particular, insects present a problem for such easy dichotomies as Western versus native.

9 The Biangai man in question always carries a machete. While upset, no one was in danger.

10 Without radio, phones, or other forms of communication, events in the United States were unknown to us. However, in the aftermath of the meeting, some concluded that Satan was active throughout the world on that day.

CHAPTER FIVE: BECOMING A MINING COMMUNITY

1 Falling in price between 2012 and 2014, gold has remained fairly strong, oscillating between $1,000 and $1,400 per ounce.

2 In 1922, Park and Nettleton found gold along Koranga and Edie Creeks. The area was declared an official goldfield in 1922, a mining warden appointed (J. H. Lukin), and a gold rush was officially declared in the press. See Waterhouse 2010 for a history of gold mining in the region.

3 But see Burton 1996b, 2000a; Moretti 2006, 2012.

4 See Gregory (1979) for a discussion of labor history in Papua New Guinea as it relates to plantations and mining.

5 E.g., Cleary 1990; Ferguson 1999; Golub 2014; Jacka 2015; Kirsch 2006, 2014; Nash 1979; Taussig 1980.

6 As Gewertz and Errington (1999) point out, "class" differences emerge only as distinctions of consumption. Such distinctions evoke what Weber (1968) called difference in status and not class.

7 Kwembu, the other Biangai community with a claim to Hidden Valley Gold Mine, had taken a further step, dividing into five separate communities.

8 Marijuana use is common throughout Papua New Guinea and readily available in the towns of Wau and Bulolo (see Halvaksz 2007; Lipset and Halvaksz 2006). The company tests employees when they are hired.

9 Niolam Catering Services was founded by landowners of the Lihir mining operations. They began operations at Hidden Valley after winning the contract in 2004.

CHAPTER SIX: WHOSE CLOSURE?

1 Serres makes the point that "people usually confuse time and the measurement of time" (Serres and Latour 1995, 60–61, italics removed). He considers Western measurements of time to be the segmented

ordering of events *as if* stretched out into a straight line. But he argues that time itself does not flow so precisely and that its reality is much more complex. Serres provides a number of metaphors: the turbulent eddies of a river (Serres 1982), the movement of the fly or the baker confronting a flat plane of dough (Serres and Latour 1995). Like the measurement of time, the distant edges of dough (or the distance that a fly travels from point A to B) can be measured by calculating the distance and time it would take to traverse the smooth line imagined between their points. However, for the baker (and the fly) linearity between points is lost in the manipulations of space; the dough must be kneaded to make bread (the fly zigzags). Like layers of sediment subjected to tectonic, volcanic, and sedimentary forces, the flat plane is folded, bringing the edges together over and over. Points that were once separated are brought into confluence, folded, and kneaded, making connections across history.

2 E.g., Hyndman 1994; Jacka 2015; Jackson 2003; Kirsch 1997, 2004, 2006.

3 By comparison, Day Dawn began with £25,000 and Bulolo Gold Dredging with US$400,000.

4 See Healy 1967; Newbury 1975; Sinclair 1998; Waterhouse 2010.

5 The Sydney Harbour Bridge was funded initially through a "betterment tax," but eventually tolls and other sources of revenue were sought (Spearritt 1982).

6 See Moretti (2006) on small-scale mining around the town of Wau.

7 Jackson argues, "Companies have, in general, shown more awareness of closure issues and done more as a result of that awareness than any other single stakeholder in PNG" (2002, 33). See also Banks 2006, 2009; Cochrane 2017; Jackson 2002.

8 See also Bourdieu 1977; Ka'ili 2017; Māhina 2010; Munn 1986.

9 See, for example, Biersack 1999; Gell 1975; Goldman 1983; Jacka 2001; Stewart and Strathern 2002.

10 See also Leach 2003; Munn 1986; Sillitoe 2003; M. Smith 1994.

CHAPTER SEVEN: BELONGING

1 There is a great deal of debate regarding the inhabitants of the San Antonio area. Coahuiltecan has been critiqued as not specific enough and as incorporating unrelated groups. Campbell and Campbell (1996) highlight how complex it is to include Tonkawa, Coahuilteco, Karankawa, Comecrudo, Cotoname, Solano, and Aranama as indigenous communities with claims to the area. In the end, this further highlights the impact of colonialism on indigenous rights.

2 See Golub 2014; Jacka 2015; and Kirsch 2006, 2014.

BIBLIOGRAPHY

Addo, Ping-Ann. 2013. *Creating a Nation with Cloth: Women, Wealth, and Tradition in the Togan Diaspora.* ASAO Studies in Pacific Anthropology. Oxford: Berghahn Books.

Advertiser (Adelaide). 1926. "New Guinea Natives Murdered: Lax Administration Alleged." December 11, 1926.

Agarwal, Arjun. 2005. *Environmentality: Technologies of Government and the Making of Subjects.* Durham: Duke University Press.

Aidwatch. 2005. *Boomerang Aid: Not Good Enough Minister! Response to Foreign Minister Alexander Downer's Comments on Boomerang Aid.* Report.

Allen, Bryant, R. Michael Bourke, and Robin Hide. 1995. "The Sustainability of Papua New Guinea Agricultural Systems: The Conceptual Background." *Global Environmental Change* 5 (4): 297–312.

Allen, Bryant, R. Michael Bourke, and Andrew McGregor. 2009. "Cash Income from Agriculture." In *Food and Agriculture in Papua New Guinea,* edited by R. Michael Bourke and Tracy Harwood, 283–424. Canberra: Australia National University Press.

Anderson, Astrid. 2011. *Landscapes of Relations and Belonging: Body, Place and Politics in Wogeo, Papua New Guinea.* Oxford: Berghahn Books.

Anderson, Benedict. 1991. *Imagined Communities: Reflections on the Origin and Spread of Nationalism.* New York: Verso.

Arens, William. 1979. *The Man-Eating Myth: Anthropology and Anthropophagy.* New York: Oxford University Press.

Argus. 1926. "New Guinea Goldfields, Richness of Edie Creek, Difficulties of Travel." August 23, 1926.

Bachelard, Gaston. 1964. *The Poetics of Space.* Translated by Maria Jolas. London: Orion Press.

Bakhtin, M. M. 1981. *The Dialogic Imagination: Four Essays.* Edited by M. Holquist. Translated by C. Emerson and M. Holquist. Austin: University of Texas Press.

Bamford, Sandra. 2007. *Biology Unmoored: Melanesian Reflections on Life and Biotechnology.* Berkeley: University of California Press.

Banks, Glenn. 1999. "Gardens and Wantoks." In *Dilemmas of Development: The Social and Economic Impact of the Porgera Gold Mine 1989–1994,* edited by Colin Filer, 160–90. Canberra: Asia-Pacific Press.

———. 2006. "Mining, Social Change and Corporate Social Responsibility: Drawing Lines in the Papua New Guinea Mud." In *Globilisation and Governance in the Pacific Islands,* edited by S. Firth, 259–74. Canberra: ANU E Press.

———. 2009. "Activities of TNCs in Extractive Industries in Asia and the Pacific: Implications for Development." *Transnational Corporations* 18 (1): 43–58.

———. 2014. *2014 National Human Development Report, Papua New Guinea.* Geneva: United Nations Development Programme.

Banks, Glenn, and Chris Ballard. 2003. "Resource Wars: The Anthropology of Mining." *Annual Review of Anthropology* 32: 287–313.

Barker, John. 2016. *Ancestral Lines: The Maisin of Papua New Guinea and the Fate of the Rainforest.* Toronto: University of Toronto Press.

Barlow, Kathleen. 1985. "Learning Cultural Meanings through Social Relationships: An Ethnography of Childhood in Murik Society, Papua New Guinea." PhD diss., University of California, San Diego.

———. 2001. "Working Mothers and the Work of Culture in a Papua New Guinea Society." *Ethos* 29 (1): 78–107.

Barlow, Kathleen, and David Lipset. 1997. "Dialogics of Material Culture: Male and Female in Murik Outrigger Canoes." *American Ethnologist* 24 (1): 4–36.

Barlow, Kathleen, and Steven Winduo, eds. 1997. "Logging the Southwestern Pacific: Perspectives from Papua New Guinea, Solomon Islands, and Vanuatu." *Contemporary Pacific* 9 (1): 1–24.

Barnes, J. A. 1962. "African Models in the New Guinea Highlands." *Man* 62: 5–9.

Bashkow, Ira. 2006. *The Meaning of Whiteman: Race and Modernity in the Orokaiva Cultural World.* Chicago: University of Chicago Press.

Beehler, Bruce, ed. 1993. *Papua New Guinea Conservation Needs Assessment.* Vol. 2, *Biodiversity Support Program and the Department of Environment and Conservation, PNG.* Landover, MD: Corporate Press.

Bell, Joshua, Paige West, and Colin Filer, eds. 2015. *Tropical Forests of Oceania: Anthropological Perspectives.* Canberra: ANU Press.

Benediktsson, Karl. 2002. *Harvesting Development: The Construction of Fresh Food Markets in Papua New Guinea.* Ann Arbor: University of Michigan Press.

Berdahl. Daphne. 2001. "'Go, Trabi, Go!': Reflections on a Car and Its Symbolization over Time." *Anthropology and Humanism* 25 (2): 131–41.

Bergson, Henri. 1950. *Matter and Memory.* Translated by Nancy Margaret Paul and W. Scott Palmer. New York: Macmillan.

———. 1971. *Time and Free Will; An Essay on the Immediate Data of Consciousness.* Translated by F. L. Pogson. New York: Humanities Press.

Bernstein, Peter L. 2000. *The Power of Gold: The History of an Obsession.* New York: John Wiley & Sons.

Berry, Wendell. (1970) 1983. *The Hidden Wound.* San Francisco: North Point Press.

Besky, Sarah, and Jonathan Padwe. 2016. "Placing Plants in Territory." *Environment and Society: Advances in Research* 7: 9–28.

Biersack, Aletta. 1999. "The Mount Kare Python and His Gold: Totemism and Ecology in the Papua New Guinea Highlands." *American Anthropologist* 101 (1): 68–87.

Bird-David, Nurit. 1999. "'Animism' Revisited: Personhood, Environment, and Relational Epistemology." *Current Anthropology* (supplement) 40 (1): 67–91.

Bissell, William. 2005. "Engaging Colonial Nostalgia." *Cultural Anthropology* 20 (2): 215–48.

Blackwood, Beatrice. 1939. "Folk-Stories of a Stone Age People in New Guinea." *Folklore* 50 (3): 209–42.

———. 1978. *The Technology of a Modern Stone Age People in New Guinea.* Oxford: Pitt Rivers Museum.

Booth, Doris. 1929. *Mountains, Gold and Cannibals.* London: Morrison and Gibb.

Boserup, Ester. 1965. *The Conditions of Agricultural Growth: The Economics of Agrarian Change under Population Pressure.* Chicago: Adeline.

Bourdieu, Pierre. 1977. *Outline of a Theory of Practice.* Translated by Richard Nice. Cambridge: Cambridge University Press.

———. 1984. *Distinction: A Social Critique of the Judgement of Taste.* Translated by R. Nice. Cambridge, MA: Harvard University Press.

———. 1991. *Language and Symbolic Power.* Cambridge: Polity Press.

Bourke, R. Michael, Bryant Allen, and Michael Lowe. 2016. *Estimated Impact of Drought and Frost on Food Supply in Rural PNG in 2015.* Policy Brief 11, January. Development Policy Centre.

Bourke, R. Michael, and Tracy Harwood, eds. 2009. *Agriculture in Papua New Guinea.* Canberra: Australia National University E-Press.

Bradley, Philip. 2008. *The Battle for Wau: New Guinea's Frontline, 1942–1943.* Cambridge: Cambridge University Press.
Brady, Ivan, ed. 1976. *Transactions in Kinship: Adoption and Fosterage in Oceania.* ASAO Monograph No. 4. Honolulu: University Press of Hawaii.
Braun, Bruce. 2002. *The Intemperate Rainforest: Nature, Culture, and Power on Canada's West Coast.* Minneapolis: University of Minnesota Press.
Brightman, Robert. 1993. *Grateful Prey: Rock Cree Human-Animal Relationships.* Berkeley: University of California Press.
Brookes, Robert. 2006. "Tracking Down the Scents of the Natural World." https://www.swissinfo.ch/eng/tracking-down-the-scents-of-the-natural-world/5611450.
Brookfield, Harold. 1972. "Intensification and Disintensification in Pacific Agriculture: A Theoretical Approach." *Pacific Viewpoint* 13: 30–48.
———. 1984. "Intensification Revisited." *Pacific Viewpoint* 13: 15–44.
Brosius, Pete. 1999. "Green Dots, Pink Hearts: Displacing Politics from the Malaysian Rainforest." *American Anthropologist* 101 (1): 36–57.
Brown, Paula, and Donald Tuzin, eds. 1983. *The Ethnography of Cannibalism.* Washington, DC: Society for Psychological Anthropology.
Bulmer, Ralph. 1967. "Why Is the Cassowary Not a Bird? A Problem of Zoological Taxonomy among the Karam of the New Guinea Highlands." *Man* 2 (1): 5–25.
Burridge, Kenelm. 1954. *Mabu: A Melanesian Millennium.* London: Methuen.
Burton, John. 1995. "What Is Best Practice? Social Issues and the Culture of the Corporation in Papua New Guinea." In *Conference Proceedings of Mining and Mineral Resource Policy Issues in Asia-Pacific Prospects for the 21st Century,* edited by D. Denoon, C. Ballard, G. Banks, and P. Hancock. Canberra: Australian National University.
———. 1996a. "Aspects of the Biangai Society: The *Solorik* System." Hidden Valley Project, Working Paper no. 3.
———. 1996b. "Early Colonial Contacts among the Upper Watut and Biangai Peoples from 1895 to the First World War." Hidden Valley Project, Working Paper no. 6.
———. 2000a. "Settlement History of the Southwestern Biangai and Catalogue of National Museum Site Code." Hidden Valley Working Paper no. 12, August 2000.
———. 2000b. "First Contacts between Outsiders and the Watut and Biangai People of the Wau and Bulolo Area." Hidden Valley Working Paper no. 13, October 2000.
———. 2003. "Fratricide and Inequality: Things Fall Apart in Eastern New Guinea." *Archaeology in Oceania* 38 (3): 203–16.

———. 2013. *Development and Social Mapping in the Hidden Valley Gold Mine Impact Area, 10 Year Re-study—Final Report.* ANU Enterprise.
Büscher, Bram, and Nico Davidov. 2013. *Uncomfortable Bedfellows? Exploring the Contradictory Natures of the Ecotourism/Extraction Nexus.* Abingdon: Routledge.
Büscher, Bram, Sian Sullivan, Katja Neves, Jim Igoe, and Dan Brockington. 2012. "Towards a Synthesized Critique of Neoliberal Biodiversity Conservation." *Capitalism Nature Socialism* 23 (2): 4–30.
Cajete, Gregory. 2000. *Native Science: Natural Laws of Interdependence.* Santa Fe, NM: Clear Light Books.
Cambrony, H. R. 1992. *Coffee Growing.* Translated by S. Barrett. London: Macmillan.
Campbell, T. N., and T. J. Campbell. 1996. *Indian Groups Associated with Spanish Missions of the San Antonio Missions National Historical Park.* San Antonio: Center for Archaeological Research, University of Texas at San Antonio.
Canberra Times. 1929. "Tropical Trade: New Guinea Resources, Overseas Interest." March 7, 1929.
———. 1938. "Wau Claims as Capital Site." September 2, 1938.
Carrier, A., and J. Carrier. 1991. *Structure and Process in a Melanesian Society: Ponam's Progress in the Twentieth Century.* Chur, Switzerland: Hardwood Academic Publishers.
Carsten, Janet. 2000. "Introduction: Cultures of Relatedness." *In Cultures of Relatedness: New Approaches to the Study of Kinship,* edited by J. Carsten, 1–36. Cambridge: Cambridge University Press.
———. 2004. *After Kinship.* Cambridge: Cambridge University Press.
Casey, Edward. 1996. "How to Get from Space to Place in a Fairly Short Stretch of Time: Phenomenological Prolegomena." In *Senses of Place,* edited by S. Feld and K. Basso, 13–52. Santa Fe, NM: School of American Research Press.
Castree, Noel. 2010. "Neoliberalism and the Biophysical Environment: A Synthesis and Evaluation of the Research." *Environment and Society: Advances in Research* 1: 5–45.
Cepek, Michael. 2008. "Essential Commitments: Identity and the Politics of Cofán Conservation." *Journal of Latin American and Caribbean Anthropology* 13 (1): 196–222.
———. 2018. *Life in Oil: Cofán Survival in the Petroleum Fields of Amazonia.* Austin: University of Texas Press.
Chand, Satish. 2003. *PNG Economic Survey: Some, Albeit Weak, Signs of Recovery.* November. https://core.ac.uk/download/pdf/156689741.pdf.

———. 2017. "Registration and Release of Customary-Land for Private Enterprise: Lessons from Papua New Guinea." *Land Use Policy* 61 (2017): 413–19.
Chayanov, Alexander. 1966. *The Theory of the Peasant Economy.* Homewood, IL: American Economic Association.
Chinnery, E. W. Pearson. 1928. *Natives of the Waria, Williams and Bialolo Watersheds.* Canberra: H. J. Green.
Chinnery, Sarah. 1998. *Malaguna Road: The Papua and New Guinea Diaries of Sarah Chinnery.* Edited and introduced by Kate Fortune. Canberra: National Library of Australia.
Chocilco (Chilean Copper Commission). 2002. *Research on Mine Closure.* Mining, Minerals and Sustainable Development Project of the International Institute for Environment and Development, and the World Business Council for Sustainable Development, Report 44 (January 2002).
Clark, Jeffrey. 1993. "Gold, Sex, and Pollution: Male Illness and Myth at Mt. Kare, Papua New Guinea." *American Anthropologist* 20 (4): 742–57.
———. 1999. "The Incredible Shrinking Big Men: Male Ideology and Development in a Southern Highlands Society." *Canberra Anthropology* 12 (1–2): 120–43.
Clarke, William. 1976. "The Maintenance of Agriculture and Human Habitats within the Tropical Forest Ecosystem." *Human Ecology* 4: 247–59.
Cleary, David. 1990. *Anatomy of the Amazon Gold Rush.* Iowa City: Iowa University Press.
Clune, Frank. 1952. *Somewhere in New Guinea.* New York: Philosophical Library.
Cochrane, Glynn. 2017. *Anthropology in the Mining Industry: Community Relations after Bougainville's Civil War.* Cham, Switzerland: Palgrave Macmillan.
Coffee Industry Newsletter. 2011. "Hidden Valley Mine Supports CIC's Objectives." Urakumpa, Eastern Highlands Province, Papua New Guinea, March 2011, 8.
Conklin, Beth, and Laura Graham. 1995. "The Shifting Middle Ground: Amazonian Indians and Eco-politics." *American Anthropologists* 97 (4): 683–94.
Conklin, Harold. 1957. *Hanunoo Agriculture in the Philippines.* Rome: Food and Agricultural Organization of the United Nations.
Coronil, Fernando. 1997. *The Magical State: Nature, Money and Modernity in Venezuela.* Chicago: University of Chicago Press.
Crapanzano, Vincent. 2004. *Imaginative Horizons: An Essay in Literarary-Philosophical Anthropology.* Chicago: University of Chicago Press.
Creed, Barbara, and Jeanette Hoorn, eds. 2001. *Body Trade: Captivity, Cannibalism and Colonialism in the Pacific.* New York: Routledge.

Daily Guardian. 1927. "Natives Killed at Bulolo: Cannibals Attack Porters on Government Works." January 14, 1927.

Davis, Susan Gray. 1997. *Spectacular Nature: Corporate Culture and the Sea World Experience*. Berkeley: University of California Press.

Deleuze, Gilles. 1988. *Bergsonism*. Minneapolis: University of Minnesota Press.

———. 1995. *Negotiations*. Translated by M. Joughin. New York: Columbia University Press.

Deleuze, Gilles, and Felix Guattari. (1980) 1987. *A Thousand Plateaus: Capitalism and Schizophrenia*. Translated by B. Massumi. Minneapolis: University of Minnesota Press.

Delle, James. 1998. *An Archaeology of Social Space: Analyzing Coffee Plantations in Jamaica's Blue Mountains*. New York: Plenum Press.

Demaitre, Edmond. 1936. *New Guinea Gold: Cannibals and Gold-seekers in New Guinea*. Boston: Houghton Mifflin.

Descola, Philippe. 1994. *In the Society of Nature: A Native Ecology in Amazonia*. Cambridge: Cambridge University Press.

Deslisle, Gilles. 2004. "A Taxonomic Revision of the 'Birdwing Butterflies of Paradise,' Genus *Ornithoptera*, based on the Adult Morphology (Lepidoptera: Papilionidae)." *Lambillionea* 104 (4): 3–93.

Dixon, Robert. 2001. "Cannibalizing Indigenous Texts: Headhunting and Fantasy in Ion L. Idriess, Coral Sea Adventures." In *Body Trade: Captivity, Cannibalism, and Colonialism in the Pacific*, edited by Barbara Creed and Jeannette Hoorn, 112–25. New York: Routledge.

Doane, Molly. 2012. *Stealing Shining Rivers: Agrarian Conflict, Market Logic, and Conservation in a Mexican Forest*. Phoenix: University of Arizona Press.

Dove, Michael. 1986. "The Practical Reason of Weeds in Indonesia: Peasant vs. State Views of Imperata and Chromolaena." *Human Ecology* 14 (2): 163–90.

Dow, D. B, J. A. J. Smit, and R. W. Page, cartographers. 1974. *Wau, Papua New Guinea*. 1:250,000 Geological Series, Sheet SB / 55-14 International Index. Canberra: Australian Government—Publishing Service.

Dubert, Raymond, and Marjorie Dubert. 1973. "Biangai Phonemes." In *Phonologies of Three Languages of Papua New Guinea*, edited by Alan Healey, 5–35. Workpapers in Papua New Guinea Languages 2. Ukarumpa: Summer Institute of Linguistics.

Dudley, Kathryn. 2002. *Debt and Dispossession: Form Loss in America's Heartland*. Chicago: University of Chicago Press.

Dutton, T. E. 1976. "Austronesian Languages: Eastern Part of South-Eastern Mainland Papua." In *New Guinea Area Languages and Language Study*, edited by S. A. Wurm, vol. 12. Pacific Linguistics Series C, no. 39.

Canberra: Department of Linguistics, Research School of Pacific Studies, Australian National University.

Edwards, Jeanette, and Marilyn Strathern. 2000. "Including Our Own." In *Cultures of Relatedness: New Approaches to the Study of Kinship*, edited by J. Carsten, 149–66. Cambridge: Cambridge University Press.

Ellis, David. 2002. "Between Custom and Biodiversity: Local Histories and Market-Based Conservation in the Pio-Tura Region of Papua New Guinea." PhD diss., University of Kent at Canterbury.

Ernst, T. 1999. "Land, Stories, and Resources: Discourse and Entification in Onabasulu Modernity." *American Anthropologist* 101 (1): 88–97.

Errington, Frederick, and Deborah Gewertz. 2004. *Yali's Question: Sugar, Culture and History*. Chicago: University of Chicago Press.

Escobar, Arturo. 1999. "After Nature: Steps to an Antiessentialist Political Ecology. *Current Anthropology* 40: 1–16.

Fajans, Jane. 1998. "Transforming Nature, Making Culture: Why Baining Are Not Environmentalists." *Social Analysis* 42 (3): 12–27.

Feit, Harvey. 1998. "Reflections on Local Knowledge and Wildlife Resource Management: Differences, Dominance and Decentralisation." In *Aboriginal Environmental Knowledge in the North*, edited by Louis-Jacques Dorais, Murielle Nagy, and Ludger Muller-Wille, 123–48. Québec: Université Laval, Gétic.

Feld, Steven. 1990. *Sound and Sentiment: Birds, Weeping, Poetics, and Song in Kaluli Expression*. Philadelphia: University of Pennsylvania Press.

Ferguson, James. 1997. "The Country and the City on the Copperbelt." In *Culture, Power, Place: Exploration in Critical Anthropology*, edited by A. Gupta and J. Ferguson, 137–54. Durham: Duke University Press.

———. 1999. *Expectations of Modernity: Myths and Meanings of Urban Life on the Zambian Copperbelt*. Berkeley: University of California Press.

Fewkes, J. Walter. 1910. "The Butterfly in Hopi Myth and Ritual." *American Anthropologist* 12 (4): 576–94.

Filer, Colin. 1990. "The Bougainville Rebellion, the Mining Industry and the Process of Social Disintegration in Papua New Guinea." *Canberra Anthropology* 13: 1–39.

———, ed. 1997. *The Political Economy of Forest Management in Papua New Guinea*. Port Moresby: National Research Institute.

———. 1998. "The Melanesian Way of Menacing the Mining Industry." In *Modern Papua New Guinea*, edited by Zimmer-Tamakoshi, 147–78. Kirksville, MO: Thomas Jefferson University Press.

———. 1999. *Dilemmas of Development: The Social and Economic Impact of the Porgera Gold Mine 1989–1994*. Canberra: Asia-Pacific Press.

Finney, Ben. 1973. *Big-men and Business, Entrepreneurship and Economic Growth in the New Guinea Highlands.* Honolulu: University Press of Hawaii.

Fisher, William. 1994. "Megadevelopment, Environmentalism, and Resistance: The Institutional Context of Kayapo Indigenous Politics in Central Brazil." *Human Organization* 53 (3): 220–32.

Fletcher, Robert. 2010. "Neoliberal Environmentality: Towards a Poststructuralist Political Ecology of the Conservation Debate." *Conservation and Society* 8 (3): 171–81.

———. 2014. *Romancing the Wild: Cultural Dimensions of Ecotourism.* Durham: Duke University Press.

Fortune, Reo. (1932) 1963. *Sorcerers of Dobu: The Social Anthropology of the Dobu Islanders of the Western Pacific.* New York: E. P. Dutton.

Foucault, Michel. 1995. *Discipline and Punishment: The Birth of the Prison.* Translated by Alan Sheridan. New York: Vintage Books.

———. 2008. *The Birth of Biopolitics: Lectures at the Collège de France, 1978–79.* London: Palgrave Macmillan.

Freeborn, Amy. 2014. "Specimen(s) of the Month #7: Butterflies with Bullet Wounds." *Nature Plus* (blog). Natural History Museum, May 30, 2014. https://www.nhm.ac.uk/natureplus/blogs/behind-the-scenes/2014/05/30/specimens-of-the-month-7-butterflies-with-bullet-wounds.html.

Frosh, Paul. 2003. *The Image Factory: Consumer Culture, Photography and the Visual Content Industry.* London: Bloomsbury.

Garvin, Theresa, Tara McGee, Karen Smoyer-Tomic, and Emmanuel Ato Aubynn. 2009. "Community-Company Relations in Gold Mining in Ghana." *Journal of Environmental Management* 90 (1): 571–86.

Geertz, Clifford. 1963. *Agricultural Involution: The Processes of Ecological Change in Indonesia.* Berkeley: University of California Press.

Gegeo, David. 2001. "Cultural Rupture and Indigeneity: The Challenge of (Re)visioning 'Place' in the Pacific." *Contemporary Pacific* 13 (2): 491–507.

Gell, Alfred. 1975. *Metamorphosis of the Cassowary: Umeda Society, Language and Ritual.* London: Athlone Press.

———. 1992. *The Anthropology of Time: Cultural Constructions of Temporal Maps and Images.* Oxford: Berg.

Gewertz, Deborah, and Frederick Errington. 1999. *Emerging Class in Papua New Guinea: The Telling of Difference.* Cambridge: Cambridge University Press.

Gibson, John. 2001. *Food Security and Food Policy in Papua New Guinea.* Port Moresby, PNG: Institute of National Affairs.

Gilberthorpe, Emma. 2013. "In the Shadow of Industry: A Study of Culturalization in Papua New Guinea." *Journal of the Royal Anthropological Institute* 19 (2): 261–78.
Givaudan. 2001. *Half Yearly Report*. Available on the Givaudan website, https://www.givaudan.com/media/corporate-publications. Accessed September 14, 2019.
———. 2004. *Analytical Chemistry*. Available on the Givaudan website, www.givaudan.com/appl/inet/gwstatic.nsf/page/se-rd-an!OpenDocument. Accessed February 14, 2005.
Godelier, Maurice, and Jose Garanger. 1979. "Stone Tools and Steel Tools among the Baruya of New Guinea: Some Ethnographic and Quantitative Data." *Social Science Information* 18: 633–78.
Godoy, Ricardo. 1985. "Mining: Anthropological Perspectives." *Annual Review of Anthropology* 14: 199–217.
———. 1990. *Mining and Agriculture in Highlands Bolivia: Ecology, History, and Commerce among the Jukumanis*. Phoenix: University of Arizona Press.
Goldman, Laurence. 1983. *Talk Never Dies: The Language of Huli Disputes*. London: Tavistock.
———, ed. 1999. "From Pot to Polemic: Uses and Abuses of Cannibalism." In *The Anthropology of Cannibalism*, edited by Laurence R. Goldman, 1–26. Westport, CT: Bergin & Harvey.
Göltenboth, Friedhelm, ed. 1990. *Subsistence Agriculture Improvement: Manual for the Humid Tropics*. Wau Ecology Institute Handbook 10. Weikersheim: Margraf Press.
Golub, Alex. 2014. *Leviathans at the Gold Mine: Creating Indigenous and Corporate Actors in Papua New Guinea*. Durham: Duke University Press.
Goodenough, Ward. 1970. "Transaction in Parenthood." In *Adoption in Eastern Oceania*, edited by V. Carroll. Honolulu: University of Hawaii Press.
Gordillo, Gaston. 2004. *Landscapes of Devils: Tensions of Place and Memory in the Argentinean Chaco*. Durham: Duke University Press.
———. 2014. *Rubble: The Afterlife of Destruction*. Durham: Duke University Press.
Graeber, David. 2015. "Radical Alterity Is Just Another Way of Saying 'Reality': A Reply to Eduardo Viverios de Castro." *Hau* 5 (2): 1–41.
Grandia, Liza. 2012. *Enclosed: Conservation, Cattle, and Commerce among the Q'eqchi' Maya Lowlanders*. Seattle: University of Washington Press.
Gregory, Chris. 1979. "The Emergence of Commodity Production in Papua New Guinea." *Journal of Contemporary Asia* 9 (4): 389–409.

Gressit, J. Linsley. 1963. *Pacific Basin Biogeography.* Honolulu: Bishop Museum Press.

Gressitt, J. L., and Nalini Nadkarni. 1978. *Guide to Mount Kaindi: Background to Montane New Guinea Ecology.* Wau, Papua New Guinea: Wau Ecology Institute.

Gudeman, Stephen. 1986. *Economics as Culture: Models and Metaphors of Livelihood.* London: Routledge and Kegan Paul.

———. 2001. *The Anthropology of the Economy: Community, Market, and Culture.* Malden, MA.: Blackwell.

———. 2016. *Anthropology and the Economy.* Cambridge: Cambridge University Press.

Guinea Airway. 1933. "A Unique Aerial Transport Service." *Pacific Islands Monthly,* March 1933.

———. 1935. "Safety–Efficiency—Economy." *Pacific Islands Monthly,* September 1935.

———. 1938a. "Building a New Civilization . . . by Air!" *Pacific Islands Monthly,* September 1938.

———. 1938b. "Time Flies . . . So does Civilization by Guinea Airways." *Pacific Islands Monthly,* November 1938.

Gupta, Akhil. 1998. *Postcolonial Developments: Agriculture in the Making of Modern India.* Durham: Duke University Press.

Halvaksz, Jamon. 1998. "Kuper Range Wildlife Management Area: Report on Fieldwork in Elauru, Morobe Province." Unpublished research report to Wau Ecology Institute, Morobe Province, Papua New Guinea.

———. 2003. "Singing about the Land among the Biangai." *Oceania* 7 (3): 153–69.

———. 2005. "Re-imagining Biangai Environments: Mining and Conservation in the Wau Bulolo Valley, Papua New Guinea." PhD diss., University of Minnesota.

———. 2006. "Becoming 'Local Tourists': Travel, Landscapes and Identity in Papua New Guinea." *Tourist Studies* 6 (2): 99–117.

———. 2007. "Cannabis and Fantasies of Development: Revaluing Relations through Land in Rural Papua New Guinea." *Australian Journal of Anthropology* 18 (1): 56–71.

———. 2008a. "Photographing Spirits: Indigenous Photography, Ancestors and the Environment in Papua New Guinea." *Visual Anthropology* 21 (4): 310–26.

———. 2008b. "Whose Closure? Appearances, Temporality and Mineral Extraction along the Upper Bulolo River, Papua New Guinea." *Journal of the Royal Anthropological Institute* 14: 21–37.

———. 2010. "The Photographic Assemblage: Duration, History and Photography in Papua New Guinea." *Anthropology and History* 21 (4): 411–29.

———. 2013a. "Mining the Forest: Epical and Novelesque Boundaries along the Upper Bulolo." In *Uncomfortable Bedfellows? Exploring the Contradictory Natures of the Ecotourism/Extraction Nexus*, edited by N. Davidov and B. Buscher. Abingdon: Routledge.

———. 2013b. "The Taste of Public Places: *Terroir* in Papua New Guinea's Emerging Nation." *Anthropological Forum* 23 (2): 142–57.

Halvaksz, Jamon, and Heather Young-Leslie. 2008. "Thinking Ecographically: Places, Ecographers, and Environmentalism." *Nature + Culture* 3 (2): 183–205.

Hammar, Lawrence. 1998. "AIDS, STDs, and Sex Work in Papua New Guinea." In *Modern Papua New Guinea*, edited by Laura Zimmer-Tamakoshi. Kirksville, MO: Thomas Jefferson University Press.

Hann, Chris. 2003. *The Postsocialist Agrarian Question: Property Relations and the Rural Condition.* Halle Studies in the Anthropology of Eurasia. Muenster: LIT Verlag.

Hansen, Karen Tranberg. 2004. "Helping or Hindering? Controversies around the International Second-hand Clothing Trade." *Anthropology Today* 20 (4): 3–9.

Haraway, Donna. 1989. *Primate Visions: Gender, Race, and Nature in the World of Modern Science.* New York: Routledge.

———. 1997. *Modest_Witness@Second_Millennium.FemaleMan©_Meets_Onco Mouse™: Feminism and Technoscience.* New York: Routledge.

———. 2003. *The Companion Species Manifesto: Dogs, People, and Significant Otherness.* Chicago: Prickly Paradigm Press.

———. 2008. *When Species Meet.* Minneapolis: University of Minnesota Press.

Harmony. 2011. *Sustainable Development Report 2011.* http://financialresults.co.za/2011/harmony_ar2011/downloads.php.

———. 2017. *Investor Brief*, December 2017. https://www.harmony.co.za/downloads/send/8-factsheets/2585-harmony-investor-brief.

Harper, Krista. 2005. "'Wild Capitalism' and 'Ecocolonialism': A Tale of Two Rivers." *American Anthropologist* 107 (2): 221–23.

Harrison, Simon. 1982. "Yams and the Symbolic Representation of Time in a Sepik River Village." *Oceania* 53: 141–62.

Harvey, David. 1982. *The Limits to Capital.* Oxford: Basil Blackwell.

———. 1996. *Justice, Nature and the Geography of Difference.* Cambridge: Blackwell.

———. 2005. *A Brief History of Neoliberalism.* Oxford: Oxford University Press.

Hau'ofa, Epli. 1993. "Our Sea of Islands." In *A New Oceania: Rediscovering Our Sea of Islands*, edited by Naidu, E. Waddle and E. Hau'ofa, 147–61. Suva: School of Social and Economic Development, University of the South Pacific.

———. 2008. *We Are the Ocean.* Honolulu: University of Hawaii Press.

Healy, Allan Michael. 1967. *Bulolo: A History of the Development of the Bulolo Region, New Guinea.* Canberra: New Guinea Research Unit, Australian National University.

Hess, Sabine. 2009. *Person and Place: Ideas, Ideals and Practice of Sociality on Vanua Lava, Vanuatu.* New York: Berghahn Books.

Hibberd, John. 1997. "The World's Largest Butterfly Still Flies in PNG." *Focus* (AusAid), March 19–22.

Hilgers, Matthieu. 2011. "The Three Anthropological Approaches to Neoliberalism." *International Social Science Journal* 61 (202): 333–523.

Himley, Matthew. 2012. "Regularizing Extractions in Andean Peru: Mining and Social Mobilization in an Age of Corporate Social Responsibility." *Antipode* 45: 2.

Hirsch, Eric. 1995. "Introduction. Landscape: Between Place and Space." In *The Anthropology of the Landscape*, edited by E. Hirsch and M. O'Hanlon, 1–30. Oxford: Oxford University Press.

———. 2004. "Environment and Economy: Mutual Connections and Diverse Perspectives." *Anthropological Theory* 4 (4): 435–53.

Horkheimer, Max, and Theodor W. Adorno. 1972. *Dialectic of Enlightenment.* New York: Herder and Herder.

Hughes, Helen. 2004. *Can Papua New Guinea Come Back from the Brink?* Sydney: Centre for Independent Studies.

Hyndman, David. 1994. *Ancestral Rain Forests and Mountains of Gold: Indigenous Peoples and Mining in New Guinea.* Boulder, CO: Westview Press.

Idriess, Ion Llewellyn. 1933. *Gold-Dust and Ashes.* Sydney: Angaus and Roberston.

Igoe, Jim. 2017. *The Nature of Spectacle: On Images, Money and Conserving Capitalism.* Phoenix: Arizona University Press.

Imbun, Benedict. 2000. "Mining Workers or 'Opportunities' Tribesmen?: A Tribal Workforce in a Papua New Guinea Mining Community." *Oceania* 71 (2): 129–49.

———. 2006. "Local Laborers in Papua New Guinea Mining: Attracted or Compelled to Work." *Contemporary Pacific: A Journal of Island Affairs* 18 (2): 315–33.

Ingold, Tim. 1980. *Hunters, Pastoralists, and Ranchers: Reindeer Economies and Their Transformations.* Cambridge: Cambridge University Press.

———. 2002. *The Perception of the Environment: Essays in Livelihood, Dwelling and Skill.* New York: Routledge.

Jacka, Jerry. 2001. "Coca-Cola and Kolo: Land, Ancestors and Development." *Anthropology Today* 17 (4): 3–8.

———. 2015. *Alchemy in the Rain Forest: Politics, Ecology, and Resilience in a New Guinea Mining Area.* Durham: Duke University Press.

———. 2018. "The Anthropology of Mining: The Social and Environmental Impacts of Resource Extraction in the Mineral Age." *Annual Review of Anthropology* 47: 61–77.

Jackson, Richard. 2002. *Capacity Building in Papua New Guinea for Community Maintenance during and after Mine Closure.* Mining, Minerals and Sustainable Development Project of the International Institute for Environment and Development, and the World Business Council for Sustainable Development. Report 181, February 2002.

———. 2003. *Muddying the Waters of the Fly: Underlying Issues or Stereotypes?* Technical Report Working Paper no.41, Resource Management in Asia-Pacific (RMAP) Program, RSPAS, ANU.

Johnston, Ana. 2017. "Becoming 'Pacific-Minded': Australian Middlebrow Writers in the 1940s and the Mobility of Texts." *Interdisciplinary Journal of Mobility Studies* 7 (1): 88–107.

Jorgensen, Dan. 1996. "Regional History and Ethnic Identity in the Hub of New Guinea: The Emergence of the Min." *Oceania* 66 (3): 189–210.

———. 2006. "Hinterland History: The Ok Tedi Mine and Its Cultural Consequences in Telefolmin." *Contemporary Pacific* 18 (2): 233–64.

Kahn, Miriam. 1990. "Stone-Faced Ancestors: The Spatial Anchoring of Myth in Wamira, Papua New Guinea." *Ethnology* 29 (1): 51–66.

Ka'ili, Tavita. 2017. *Marking Indigeneity.* Phoenix: University of Arizona Press.

Ka'ili, Tavita, Ōkusitino Māhina, and Ping Ann Addo, eds. 2017. "Ta-Va (Time-Space) Theory of Reality: The Birth of an Indigenous Moana Theory." *Pacific Studies Journal* 40 (1–2): 1–17.

Keesing, Roger M. 1992. *Custom and Confrontation: The Kwaoi Struggle for Cultural Autonomy.* Chicago: University of Chicago Press.

Kirksey, Eben. 2012. *Freedom in Entangled Worlds: West Papua and the Architecture of Global Power.* Durham: Duke University Press.

———. 2015. *Emergent Ecologies.* Durham: Duke University Press.

Kirsch, Stuart. 1997. "Indigenous Response to Environmental Impact along the Ok Tedi." In *Compensation for Resource Development in Papua New Guinea*, edited by S. Toft, 143–55. Boroko, Papua New Guinea: Law Reform Commission.

———. 2004. "Changing Views of Place and Time along the Ok Tedi." In *Mining and Indigenous Lifeworlds in Australia and Papua New Guinea*, edited by A. Rumsey and J. Weiner, 182–207. Canon Pyon, UK: Sean Kingston.

———. 2006. *Reverse Anthropology: Indigenous Analysis of Social and Environmental Relations in New Guinea*. Stanford, CA: Stanford University Press.

———. 2014. *Mining Capitalism: The Relationship between Corporations and Their Critics*. Oakland: University of California Press.

Knight, John. 2003. *Waiting for Wolves in Japan: An Anthropological Study of People-Wildlife Relations*. Oxford University Press.

Kohn, Eduardo. 2013. *How Forests Think: Toward an Anthropology Beyond the Human*. Berkeley: University of California Press.

Kosek, Jake. 2006. *Understories: The Political Life of Forests in Northern New Mexico*. Durham: Duke University Press.

Latour, Bruno. 1993. *We Have Never Been Modern*. Cambridge, MA: Harvard University Press.

———. 1999. *Pandora's Hope: Essays on the Reality of Science Studies*. Cambridge, MA: Harvard University Press.

———. 2004. *Politics of Nature: How to Bring the Sciences into Democracy*. Translated by Catherine Porter. Cambridge, MA: Harvard University Press.

———. 2005. *Reassembling the Social: An Introduction to Actor-Network-Theory*. Oxford: Oxford University Press.

Lawrence, Peter. (1964) 1979. *Road Belong Cargo*. New Jersey: Humanities Press.

Lea, David, and Timothy Curtin. 2011. *Land Law and Economic Development in Papua New Guinea*. Cambridge: Cambridge Scholars Publishing.

Leach, James. 2003. *Creative Land: Place and Procreation on the Rai Coast of Papua New Guinea*. New York: Berghahn Books.

Lebot, Vincent. 2009. *Tropical Root and Tuber Crops: Cassava, Sweet Potato, Yams and Aroids*. Wallingford, UK: CABI.

Lefebvre, Henri. 1992. *The Production of Space*. New York: Wiley-Blackwell.

Levi-Strauss, Claude. 1963. *Totemism*. Translated by Rodney Needham. Boston: Beacon Press.

Li, Fabiana. 2015. *Unearthing Conflict: Corporate Mining, Activism, Expertise in Peru*. Durham: Duke University Press.

Li, Tania Murray. 2007. *The Will to Improve: Governmentality, Development, and the Practice of Politics*. Durham: Duke University Press.

———. 2014. *Land's End: Capitalist Relations on an Indigenous Frontier*. Durham: Duke University Press.

Lilomaiava-Doktor, Sa'iliemanu. 2009. "Beyond 'Migration': Samoan Population Movement *(Malaga)* and the Geography of Social Space (Va)." *Contemporary Pacific* 21 (1): 1–32.

Lindenbaum, Shirley. 2004. "Thinking about Cannibalism." *Annual Review of Anthropology* 33: 475–98.

Lipset, David. 1997. *Mangrove Man: Dialogics of Culture in the Sepik Estuary.* Cambridge: Cambridge University Press.

———. 2017. "Masculinity and the Culture of Rising Sea-Levels in a Mangrove Lagoon in Papua New Guinea." *Maritime Studies* 16 (2). https://doi.org/10.1186/s40152-017-0057-5.

Lipset, David, and Jamon Halvaksz, eds. 2006. "Marijuana in Papua New Guinea." Special issue, *Oceania* 76 (3).

Losey, John E., and Mace Vaughan. 2006. "The Economic Value of Ecological Services Provided by Insects." *Bioscience* 56 (4): 311–23.

Machio, Thomas. 1994. *To Remember the Faces of the Dead: The Plentitude of Memory in Southwestern New Britain.* Madison: University of Wisconsin Press.

Mackintosh, Maureen. 1989. *Gender, Class and Rural Transition: Agribusiness and the Food Crisis in Senegal.* London: Zed Books.

Māhina, Ōkusitino. 2010. "Tā, Vā, and Moana: Temporality, Spatiality, and Indigeneity." *Pacific Studies* 33 (2): 168–202.

———. 2017. "Time, Space, and Culture: A New Ta-Va Theory of Moana Anthropology." *Pacific Studies* 40 (1–2): 169–202.

Malandina, Bernard. 1985. "Wau Landowners Hand Out a Two-Week Ultimatum." *Times of Papua New Guinea*, June 2, 1985.

Martin, Jean-Claude. 1992. "Le changement social et la question de la resistance aux techniques nouvelles chez les Biangai de Papouasie, Nouvelle-Guinea." PhD diss., University of Montreal.

Martin, Jean-Claude, and Françoise-Romaine Ouellette. 1981. "Golden Future? The Dilemma of the Biangais of the Wau Valley, Morobe Province." *Cultural Survival* 7: 50–58.

Marx, Karl. 1979. *Capital: A Critique of Political Economy.* Vol. 1. New York: International Publishers.

Maschio, T. 1994. *To Remember the Faces of the Dead: The Plentitude of Memory in Southwestern New Britain.* Madison: University of Wisconsin Press.

Mauss, Marcel. (1925) 1990. *The Gift: The Form and Reason for Exchange in Archaic Societies.* Translated by D. D. Halls. New York: W. W. Norton.

McArthur, A. Margaret. 2000. *The Curbing of Anarchy in Kunimaipa Society.* Edited by Douglas Oliver. Oceania Monography 19. Sydney: University of Sydney.

McElhanon, Kenneth. 1984. *A Linguistic Field Guide to the Morobe Province, Papua New Guinea.* Canberra: Dept. of Linguistics, Research School of Pacific Studies, Australian National University.

Mead, Margaret. 1935. *Sex and Temperament in Three Primitive Societies.* London: Penguin.

———. 1938. "The Mountain Arapesh I: An Importing Culture." *Anthropological Papers of the American Museum of Natural History* 36 (3): 139–349.

Meillassoux, Claude. 1975. *Maidens, Meal and Money: Capitalism and the Domestic Community.* Cambridge: Cambridge University Press.

Melbourne Tribune. 1926. "Rivers of Gold, Winning Wealth Mid Tropic Ferns, Headhunters, and Fearsome Snakes." October 16, 1926.

Mitchell, Timothy. 2002. *Rule of Experts: Egypt, Techno-politics, Modernity.* Berkeley: University of California Press.

Mitio, Ngawae. 1981. "Biangai Marriage and Its Relationship to Kinship and Property: The Case of Werewere Village, Wau District." Honours subthesis, University of Papua New Guinea, Port Moresby.

———. 1984. *Factors Affecting People's Attitudes to Company Exploitation of Their Resources: The Case of Timber.* Boroko, PNG: Institute of Applied Social and Economic Research.

Moore, Donald. 2005. *Suffering for Territory: Race, Place, and Power in Zimbabwe.* Durham: Duke University Press.

Moretti, Daniele. 2006. "The Gender of Gold: An Ethnographic and Historical Account of Women's Involvement in Artisanal and Small-Scale Mining in Mount Kaindi, Papua New Guinea." *Oceania* 76 (2): 133–49.

———. 2012. "Gold, Tadpoles and Jesus in the Manger: Mythopoeia, Colonialism and Redress in the Morobe Goldfields in Papua New Guinea." *Journal of the Polynesian Society* 121 (2): 151–79.

Morgan, Lewis Henry. 1871. *Systems of Consanguinity and Affinity of the Human Family.* Washington, DC: Smithsonian Institution.

Morobe Miner. 2011. "Watut Farmers Attend NARI Innovations Show." August 2011, 9.

———. 2012. "Locals Receive Business Skills Training." March 2012, 8.

Mosko, Mark. 2001. "Syncretic Persons: Sociality, Agency and Personhood in Recent Charismatic Ritual Practices among North Mekeo (PNG)." *Australian Journal of Anthropology* 12 (3): 259–74.

Munn, Nancy. 1986. *The Fame of Gawa: A Symbolic Study of Value Transformation in a Massim (Papua New Guinea) Society.* Cambridge: Cambridge University Press.

———. 1992. "The Cultural Anthropology of Time: A Critical Essay." *Annual Review of Anthropology* 21: 93–123.

Nabobo-Baba, Unaisi. 2006. *Knowing and Learning: An Indigenous Fijian Approach.* Suva: University of South Pacific Press.

Narotzky, Susana. 2004. "Provisioning." In *A Handbook of Economic Anthropology*, edited by J. Carrier, 78–93. Cheltenham: Edward Elgar.

Nash, Jill, and Eugene Ogan. 1990. "The Red and the Black: Bougainvillean Perceptions of Other Papua New Guineans." *Pacific Studies* 13: 1–17.

Nash, June. 1979. *We Eat the Mines and the Mines Eat Us: Dependency and Exploitation in Bolivian Tin Mines.* New York: Columbia University Press.

———. 1992. *I Spent My Life in the Mines: The Story of Juan Rojas, Bolivian Tin Miner.* New York: Columbia University Press.

National Census Office, Papua New Guinea. 2014. *2011 National Population and Housing Census, Ward Population Profile, Momase Region.* Port Moresby, PNG: National Statistical Office, National Capital District.

Nelson, Hank. 1976. *Black White and Gold: Goldmining in Papua New Guinea, 1878–1930.* Canberra: Australian National University Press.

Netting, Robert. 1974. "Agrarian Ecology." *Annual Review of Anthropology* 3: 21–56.

———. 1993. *Smallholders, Householders: Farm Families and the Ecology of Intensive Sustainable Agriculture.* Stanford, CA: Stanford University Press.

New, T. R., R. M. Pyle, J. A. Thomas, C. D. Thomas, and P. C. Hammond. 1995. "Butterfly Conservation Management." *Annual Review of Entomology* 40: 57–83.

Newbury, Colin. 1975. "Colour Bar and Labour Conflict on the New Guinea Goldfields 1935–1941." *Australian Journal of Politics and History* 21 (3): 25–58.

Nordhagen, Stella, Unai Pascual, and Adam Drucker. 2017. "Feeding the Household, Growing the Business, or Just Showing Off? Farmers' Motivations for Crop Diversity Choices in Papua New Guinea." *Ecological Economics* 137: 99–109.

Obeyesekere, Gananath. 1992. "'British Cannibals': Contemplation of an Event in the Death and Resurrection of James Cook, Explorer." *Critical Inquiry* 18 (Summer): 630–54.

———. 1998. "Cannibal Feasts in Nineteenth-Century Fiji: Seamen's Yarns and the Ethnographic Imagination." In *Cannibalism and the Colonial World*, edited by Francis Barker, Peter Hulme, and Margaret Iversen, 63–86. Cambridge: University of Cambridge Press.

———. 2001. "Narratives of the Self: Chevalier Peter Dillon's Fijian Cannibal Adventures." In *Body Trade: Captivity, Cannibalism and Colonialism in the Pacific*, edited by Barbara Creed and Jeannette Hoorn, 69–111. New York: Routledge.

Ogan, Eugene. 1996. "Copra Came before Copper: The Nasioi of Bougainville and Plantation Colonialism, 1902–1964." *Pacific Studies* 19: 31–51.
O'Hanlon, Michael, and Linda Frankland. 2003. "Co-present Landscapes: Routes and Rootedness as Sources of Identity in Highlands New Guinea." In *Landscape, Memory and History: Anthropological Perspectives*, edited by P. Stewart and A. Strathern, 166–88. London: Pluto Press.
O'Neil, Jack. 1979. *Up from South: A Prospector in New Guinea 1931–1937*. Edited by J. Sinclair. Oxford: Oxford University Press.
Orsak, Larry. 1993. "Farming Butterflies to Save the Rainforest." *News Lepid. Soc.* 3: 71–80.
Ortner, Sherry. 1974. "Is Female to Male as Nature Is to Culture?" In *Woman, Culture, and Society*, edited by M. Rosaldo and L. Lamphere, 68–87. Stanford, CA: Stanford University Press.
Ouellette, Françoise-Romaine. 1987. "Rapport de sexe et developpement capitaliste chez les Biangai de la region de Wau." PhD diss., University of Montreal.
Pacific Islands Monthly. 1931. "Why New Guinea Goldfields Are Producing No Gold." December 18, 1931.
———. 1932a. "N.G. Goldfields Ltd. New Accounts Show Co.'s Position—2nd Annual Report." February 22, 1932.
———. 1932b. "Gold Search and Discoveries in Pacific Islands." November 23, 1932.
———. 1933a. "New Guinea Goldfields, LTD: Annual Accounts Show Company at Interesting Stage of Development." March 23, 1933.
———. 1933b. "N.G. Goldfields Shareholders as for Information." May 18, 1933.
———. 1933c. "Gold at £7/15/- per oz Is Stimulating Mining Companies." July 19, 1933.
———. 1934. "By Air or by Road? A Problem to Be Solved by N.G. Miner." August 14, 1934.
———. 1936. "Kaisinik: Model Village in Place of Cannibal Den." October 10, 1936.
———. 1950. "Death of C. L. B. Wilde." March 1950.
Papuan Courier. 1938. "Wau as Capital." September 9, 1938.
Parsons, M. 1992. "Butterfly Farming and Conservation in the Indo-Australian Region." *Tropical Lepidoptera* 3 (Supplement): 1–31.
Peck, Jamie, and Adam Tickell. 2002. "Neoliberalizing Space." *Antipode* 34 (3): 380–404.
Perz, Stephen. 2001. "Household Demographic Factors as Life Cycle Determinants of Land Use in the Amazon." *Population Research and Policy Review* 20: 159–86.

Pictorial Sun. 1927. "Natives Shot: Arrow Attacks in New Guinea, Vindictive Tribe." January 22, 1927.

Pini, Barbara, Robyn Mayes, and Paula McDonald. 2010. "The Emotional Geography of a Mine Closure: A Study of the Ravensthorpe Nickel Mine in Western Australia." *Social and Cultural Geography* 11 (6): 559–74.

Polanyi, Karl. 1944. *The Great Transformation.* New York: Rinehart.

Post-Courier (Papua New Guinea). 1980. "Wau Group Denied K17m Sum." October 24, 1980.

———. 1998. "Mining Ruined Farming Land, Says Tribal Compensation Claim." November 5, 1998.

———. 2000. "Canadian Court Convicts PNG Butterfly Thief." December 12, 2000.

Powell, Dana. 2018. *Landscapes of Power: Politics of Energy in the Navajo Nation.* Durham: Duke University Press.

Pratt, Mary Louise. 1992. *Imperial Eyes: Travel Writing and Transculturation.* New York: Routledge.

Raffles, Hugh. 2002. *In Amazonia: A Natural History.* Princeton, NJ: Princeton University Press.

Reed, Adam. 2003. *Papua New Guinea's Last Place: Experiences of Constraint in a Post-colonial Prison.* New York: Berghahn Books.

Refiti, Albert. 2017. "How the Ta-Va Theory of Reality Constructs a Spatial Exposition of Samoan Architecture." *Pacific Studies* 40 (1–2): 267–88.

Register (Adelaide). 1926a. "Mining Note: Bulolo Goldfields." November 24, 1926.

———. 1926b. "Life in New Guinea: The New Gold Rush." October 14, 1926.

———. 1927. "New Guinea Natives: Attack on Warden's Party. A Few Casualties." January 22, 1927.

———. 1928. "Guinea Gold: Mrs. Booth of Bulolo." May 13, 1928.

Rhys, Lloyd. 1942. *High Lights and Flights in New Guinea: Being in the Main an Account of the Discovery and Development of the Morobe Goldfields.* London: Hodder and Stoughton.

Richards, A. 1939. *Land, Labour and Diet in Northern Rhodesia.* London: Oxford University Press.

Ritvo, Harriet. 1997. *The Animal Estate: The English and Other Creatures in the Victorian Age.* Cambridge, MA: Harvard University Press.

Rivers, W. H. R. 1900. "A Genealogical Method of Collecting Social and Vital Statistics." *Journal of the Royal Anthropological Institute* 30: 74–82.

———. 1910. "The Genealogical Method of Anthropological Enquiry." *Sociological Review* 3: 1–12.

Robbins, Joel. 2004. *Becoming Sinners: Christianity and Moral Torment in a Papua New Guinea Society.* Berkeley: University of California Press.

Robbins, Paul. 2007. *Lawn People: How Grasses, Weeds, and Chemicals Make Us Who We Are.* Philadelphia: Temple University Press.

Roberts, Mere, Brad Haami, Richard Benton, Terre Satterfield, Melissa L Finucane, Mark Henare, and Manuka Henare. 2004. "Whakapapa as Maori Mental Construct: Some Implications for the Debate over Genetic Modification of Organisms." *Contemporary Pacific* 16 (1): 1–28.

Rodman, Margaret Critchlow. 1987. *Masters of Tradition: Consequences of Customary Land Tenure in Langana, Vanuatu.* Vancouver: University of British Columbia Press.

Rosaldo, Michelle Zimbalist. 1980. *Knowledge and Passion: Ilongot Notions of Self and Social Life.* Cambridge: Cambridge University Press.

Rumsey, Alan. 1999. "Social Segmentation, Voting, and Violence in Papua New Guinea." *Contemporary Pacific* 11 (2): 305–33.

Rumsey, Alan, and James Weiner, eds. 2004. *Mining and Indigenous Life-worlds in Australia and Papua New Guinea.* Wantage, UK: Sean Kingston Publishing.

Sahlins, Marshall. 1985. *Islands of History.* Chicago: University of Chicago Press.

———. 2003. "Artificially Maintained Controversies: Global Warming and Fijian Cannibalism." *Anthropology Today* 19 (3): 3–5.

Sakulas, Harry. 1998. *Wau Ecology Institute, 25 Years On.* Wau, Papua New Guinea: Wau Ecology Institute.

Sakulas, H., J. Bauer, J. Birckhead, and T. de Lacy. 2000. "Integrated Conservation and Development in Papua New Guinea: Some Lessons from Wau Ecology Institute." In *Conservation in Production Environments: Managing the Matrix,* edited by J. L. Craig, N. Mitchell, and D. Saunders, 90–102. Chipping Norton, NSW: Surrey Beatty & Sons.

Sakulas, Harry, and Lawrence Tjame. 1991. *Ecological Damage Caused by the Discharges from the Ok Tedi Copper and Gold Mine, Papua New Guinea: Case Document on the Ok Tedi Mine.* Wau, Papua New Guinea: Wau Ecology Institute.

Samways, Michael. 2005. *Insect Diversity Conservation.* Cambridge: Cambridge University Press.

Sawyer, Suzana. 2004. *Crud Chronicles: Indigenous Politics, Multinational Oil, and Neoliberalism in Ecuador.* Durham: Duke University Press.

Schneider, David. 1965. "Kinship and Biology." In *Aspects of the Analysis of Family Structure,* edited by H. J. Coale, 83–101. Princeton, NJ: Princeton University Press.

———. 1984. *A Critique of the Study of Kinship*. Ann Arbor: University of Michigan.
Scott, James. 1999. *Seeing Like a State: How Certain Schemes to Improve the Human Condition Have Failed*. New Haven, CT: Yale University Press.
Sekhran, N., and S. Miller, eds. 1995. *Papua New Guinea Country Study on Biological Diversity*. Prepared for the Department of Environment and Conservation, Conservation Resource Centre, and the African Centre for Resources and Environment. Hong Kong: Colorcraft.
Serres, Michel. 1968. *Le système de Leibniz et ses modeles mathematiques*. Paris: Presses Universitaires de France.
———. 1982. *Hermes: Literature, Science, Philosophy*. Edited and translated by J. Harari and D. Bell. Baltimore, MD: Johns Hopkins University Press.
———. 2009. *The Five Senses: A Philosophy of Mingled Bodies*. London: Bloomsbury Academic.
Serres, Michel, and Bruno Latour. 1995. *Conversations on Science, Culture, and Time*, Translated by R. Lapidus. Ann Arbor: University of Michigan Press.
Sillitoe, Paul. 2002. "Contested Knowledge, Contingent Classification: Animals in the Highlands of Papua New Guinea." *American Anthropologist* 104 (4): 1162–71.
———. 2003. *Managing Animals in New Guinea*. London: Routledge.
———. 2010. *From Land to Mouth: The Agriculture "Economy" of the Wola of the New Guinea Highlands*. New Haven, CT: Yale University Press.
Simmel, Georg. 1971. "The Metropolis and Mental Life." In *Georg Simmel on Individuality and Social Forms*, edited by Donald Levine, 324–49. Chicago: Chicago University Press.
———. 1978. *The Philosophy of Money*. London: Routledge and Kegan Paul.
Sinclair, James. 1978. *Wings of Gold: How the Aeroplane Developed New Guinea*. Sydney: Pacific Publications.
———. 1998. *Golden Gateway: Lae and the Province of Morobe*. Bathurst, NSW: Crawford House Publishing.
Small, Rob. 2007. "Becoming Unstainable? Recent Trends in the Formal Sector of Insect Trading in Papua New Guinea." *Oryx* 41 (3): 386–89.
Smith, Linda Tuhiwai. 1999. *Decolonizing Methodologies: Research and Indigenous Peoples*. London: Zed Books.
Smith, Michael. 1994. *Hard Times on Kairiru Island: Poverty, Development, and Morality in a Papua New Guinea Village*. Honolulu: University of Hawaii Press.
Spearritt, Peter. 1982. *The Sydney Harbour Bridge*. London: Allen and Unwin.
Stella, Regis. 2007. *Imagining the Other: The Representation of the Papua New Guinean Subject*. Honolulu: University of Hawaii Press.

Stewart, Pamela, and Andrew Strathern. 2002. *Remaking the World: Myth, Mining and Ritual Change and the Duna of Papua New Guinea*. Washington, DC: Smithsonian Institution Press.

Strathern, Andrew. 1971. *The Rope of Moka: Ceremonial Exchange in Mount Hagen New Guinea*. Cambridge: Cambridge University Press.

———. 1982. "The Division of Labor and Processes of Social Change in Mount Hagen." *American Ethnologist* 9: 307–19.

———. 1984. *A Line of Power.* London: Tavistok.

———. 1987. "Social Classes in Mount Hagen? The Early Evidence." *Ethnology* 26 (4): 245–60.

Strathern, Andrew, and P. Stewart. 2000. *Arrow Talk: Transaction, Transition, and Contradiction in New Guinea Highlands History.* Kent, OH: Kent State University Press.

Strathern, Andrew, and Gabriele Stürzenhofecker, eds. 1994. *Migration and Transformations: Regional Perspectives.* Pittsburgh: University of Pittsburgh Press.

Strathern, Marilyn. 1980. "No Nature, No Culture: The Hagen Case." In *Nature, Culture and Gender,* edited by C. MacCormack and M. Strathern, 174–222. Cambridge: Cambridge University Press.

———. 1990. *The Gender of the Gift.* Berkeley: University of California Press.

———. 1996. "Cutting the Network." *Journal of the Royal Anthropological Institute* 2: 517–35.

Strauss, Jennifer. 1998. "Literary Culture 1914–1939: Battlers All." In *The Oxford Literary History of Australia,* edited by Bruce Bennett and Jennifer Strauss, 107–29. Melbourne: Oxford University Press.

Sunday Mail (Adelaide). 1926. "New El Dorado, Gold Discoveries in New Guinea. Field Immensely Rich, but Limited." August 28, 1926.

Sydney Sun. 1926a. "Gold Rush, New Guinea Boom, 150 Prospectors." July 8, 1926.

———. 1926b. "Gold Stacks, Riches of New Guinea Field, Prospector's Stories." July 20, 1926.

Sykes, Karen. 2001. "Paying a School Fee Is a Father's Duty: Critical Citizenship in Central New Ireland." *American Ethnologist* 28 (1): 5–31.

TallBear, Kim. 2014. "Standing With and Speaking as Faith: A Feminist-Indigenous Approach to Inquiry." *Journal of Research Practice* 10 (2). http://jrp.icaap.org/index.php/jrp/article/view/405/371.

Taussig, Michael. 1980. *The Devil and Commodity Fetishism in South America.* Chapel Hill: University of North Carolina Press.

Taylor, John Patrick. 2008. *The Other Side: Ways of Being and Place in Vanuatu.* Honolulu: University Hawaii Press.

Teaiwa, Katerina. 2015. *Consuming Ocean Island: Stories of People and Phosphate from Banaba*. Bloomington: Indiana University Press.
Tengan, Ty. 2008. *Native Men Remade: Gender and Nation in Contemporary Hawai'i*. Durham: Duke University Press.
Thoreau, Henry David. 1854. *Walden, or Life in the Woods*. Boston: Houghton, Mifflin.
Thornton, Thomas. 2015. *Being and Place among the Tlingit*. Seattle: University of Washington Press.
Todd, Zoe. 2016. "An Indigenous Feminist's Take on the Ontological Turn: 'Ontology' Is Just Another Word for Colonialism." *Journal of Historical Sociology* 29 (1): 4–22.
Trubek, Amy. 2008. *The Taste of Place: A Cultural Journey into Terroir*. Berkeley: University of California Press.
Tsing, Anna Lowenhaupt. 2000. "Inside the Economy of Appearances." *Public Culture* 12 (1): 115–44.
———. 2005. *Friction: An Ethnography of Global Connection*. Princeton, NJ: Princeton University Press.
———. 2015. *The Mushroom at the End of the World: On the Possibilities of Life in Capitalist Ruins*. Princeton, NJ: Princeton University Press.
Tuan, Yi-Fu. 1977. *Space and Place: The Perspective of Experience*. Minneapolis: University of Minnesota Press.
Tuwere, Ilaita. 2002. *Vanua: Towards a Fijian Theology of Place*. Suva, Fiji: University of the South Pacific.
Tuzin, Donald. 1972. "Yam Symbolism in the Sepik: An Interpretative Account." *Southwestern Journal of Anthropology* 28 (3): 230–54.
Udvardy, Monica. 1995. "The Lifecourse of Property and Personhood: Provisional Women and Enduring Men among the Giriama of Kenya." *Research in Economic Anthropology* 16: 325–48.
van Helden, Flip. 2001. "Through the Thicket: Disentangling the Social Dynamics of an Integrated Conservation and Development Project on Mainland Papua New Guinea." PhD diss., Rural Development Sociology, Wageningen University.
Vayda, Andrew, and Bradley Walters. 1999. "Against Political Ecology." *Human Ecology* 27 (1): 167–79.
Viveiros de Castro, Eduardo. 2004. "Perspectival Anthropology and the Method of Controlled Equivocation." *Tipití: Journal of the Society for the Anthropology of Lowland South America* 2 (1): 3–20.
Waddell, Eric. 1972. *The Mound Builders: Agricultural Practices, Environment, and Society in the Central Highlands of New Guinea*. Seattle: University of Washington Press.

Wagner, John. 2002. "Commons in Transition: An Analysis of Social and Ecological Change in a Coastal Rainforest Environment in Rural Papua New Guinea." PhD diss., McGill University, Montreal.

———. 2005. "The Politics of Accountability: An Institutional Analysis of the Conservation Movement in Papua New Guinea." In *Anthropology and Consultancy: Issues and Debates*, edited by Pamela J. Stewart and Andrew Strathern, 88–105. New York: Berghahn Books.

———. 2007. "Conservation as Development in Papua New Guinea: The View from Blue Mountain." In *Human Organization* 66 (1): 28–37.

Wagner, Roy. 1967. *The Curse of Souw: Principles of Daribi Clan Definition and Alliance in New Guinea.* Chicago: University of Chicago Press.

———. 1975. *The Invention of Culture.* Englewood Cliffs, NJ: Prentice-Hall.

Walsh, Andrew. 2012. *Made in Madagascar: Sapphires, Ecotourism, and the Global Bazaar.* Toronto: University of Toronto Press.

Walter, M. A. H. B. 1983. "Strategic Gardening." *Oceania* 53: 389–99.

Wardlow, Holly. 2006. *Wayward Women: Sexuality and Agency in a New Guinea Society.* Berkeley: University of California Press.

Wassmann, Jurg. 2001. "The Politics of Religious Secrecy." In *Emplaced Myth: Space, Narrative, and Knowledge in Aboriginal Australia and Papua New Guinea*, edited by A. Rumsey and J. Weiner, 43–70. Honolulu: University of Hawaii Press.

Waterhouse, Michael. 2010. *Not a Poor Man's Field: The New Guinea Goldfields to 1942: An Australian Colonial History.* Sydney: Halstead Press.

Watson, J. B. 1967. "Horticultural Traditions in the Eastern New Guinea Highlands." *Oceania* 38 (2): 81–98.

Watts, Vanessa. 2013. "Indigenous Place-Thought and Agency amongst Humans and Non-humans (First Woman and Sky Woman Go on a European World Tour!)." *Decolonization: Indigeneity, Education and Society* 2 (1): 20–34.

Wau Ecology Institute. 1974. *Biannual Report.* Wau, PNG: The Institute.

———. 1976. *Biannual Report.* Wau, PNG: The Institute.

Weber, Max. 1930. *The Protestant Ethic and the Spirit of Capitalism.* Translated by Talcott Parsons. London: Unwin Hyman.

———. 1968. *On Charisma and Institute Building: Selected Papers.* Edited by S. N. Eisenstadt. Chicago: Chicago University Press.

Weiner, Annette. 1976. *Women of Value, Men of Renown: New Perspectives in Trobriand Exchange.* Austin: University of Texas Press.

———. 1992. *Inalienable Possessions: The Paradox of Keeping-While-Giving.* Berkeley: University of California Press.

Weiner, James. 1988. *The Heart of the Pearl Shell: The Mythological Dimenstion of Foi Sociality.* Berkeley: University of California Press.

———. 1991. *The Empty Place: Poetry, Space, and Being among the Foi of Papua New Guinea.* Bloomington: University of Indiana Press.

———. 1998. "Revealing the Grounds of the Life in Papua New Guinea." *Social Analysis* 42 (3): 135–42.

———. 2004. "Introduction: Depositings." In *Mining and Indigenous Lifeworlds in Australia and Papua New Guinea*, edited by Alan Rumsey and James Weiner. Wantage, UK: Sean Kingston Publishing.

Welker, Marina. 2009. "Corporate Security Begins in the Community: Mining, the Corporate Social Responsibility Industry, and Environmental Advocacy in Indonesia." *Cultural Anthropology* 24 (1): 142–79.

———. 2014. *Enacting the Corporation: An American Mining Firm in Postauthoritarian Indonesia.* Berkeley: University of California Press.

Wells, Charles Valentine. 1939. Letter to R. W. Robson, November 20, 1938. Australian Archives, Pacific Manuscript Bureau 364.

Wendt, Albert. 1999. "Afterword: Tatauing the Post-colonial Body." In *Inside Out: Literature, Cultural Politics and Identity in the New Pacific*, edited by Vilsoni Hereniko and Rob Wilson, 399–412. Lanham, MD: Rowman and Littlefield.

West, Paige. 2006. *Conservation Is Our Government Now: The Politics of Ecology in Papua New Guinea.* Durham: Duke University Press.

———. 2008. "Scientific Tourism: Imagining, Experiencing, and Portraying Environment and Society in Papua New Guinea." *Current Anthropology* 49 (4): 597–626.

———. 2012. *From Modern Production to Imagined Primitive: The Social World of Coffee from Papua New Guinea.* Durham: Duke University Press.

———. 2016. *Dispossession and the Environment: Rhetoric and Inequality in Papua New Guinea.* New York: Columbia University Press.

Willis, Ian. 1977a. "A New Guinea Outrage: The Killings at Kaisenik 1926–1927 (Part I)." *Journal of the Morobe Province Historical Society* 4 (1): 2–10.

———. 1977b. "A New Guinea Outrage: The Killings at Kaisenik 1926–1927 (Part II)." *Journal of the Morobe Province Historical Society* 4 (2): 7–16.

Winduo, Steven. 2000. *Hembemba: Rivers of the Forest.* Port Moresby, PNG: Language and Literature Department, University of Papua New Guinea.

Wittman, Hannah. 2011. "Food Sovereignty: A New Rights Framework for Food and Nature?" *Environment and Society: Advances in Research* 2 (1): 87–105.

Wolf-Meyer, Matthew. 2012. *The Slumbering Masses: Sleep, Medicine, and Modern American Life.* Minneapolis: University of Minnesota Press.

Wood, Briar, and Steven Winduo. 2006. "In Spirit's Voices: An Interview with Steven Winduo." *Journal of Postcolonial Writing* 42 (1): 84–93.
Wooley, Ellen, and Friedhelm Göltenboth. 1991. *Handbook of the Wau Ecology Institute*. Vol. 11, *Medicinal Plants of Papua New Guinea*. Weikersheim: Margraf Press.
Wordie, J. R. 1983. "The Chronology of English Enclosure, 1500–1914." *Economic History Review* 36 (4): 483–505.
Worsley, Peter. 1968. *The Trumpet Shall Sound*. New York: Schoken Books.
Young-Leslie, Heather. 2007. "A Fishy Romance; Ecography and the Geopolitics of Desire." *Contemporary Pacific* 19 (2): 365–408.

INDEX

C

D

E

I

J

K

N

O

P

Y

CULTURE, PLACE, AND NATURE

Studies in Anthropology and Environment

Sacred Cows and Chicken Manchurian: The Everyday Politics of Eating Meat in India, by James Staples

Gardens of Gold: Place-Making in Papua New Guinea, by Jamon Alex Halvaksz

Shifting Livelihoods: Gold Mining and Subsistence in the Chocó, Colombia, by Daniel G. L. Tubb

Disturbed Forests, Fragmented Memories: Jarai and Other Lives in the Cambodian Highlands, by Jonathan Padwe

The Snow Leopard and the Goat: Politics of Conservation in the Western Himalayas, by Shafqat Hussain

Roses from Kenya: Labor, Environment, and the Global Trade in Cut Flowers, by Megan A. Styles

Working with the Ancestors: Mana *and Place in the Marquesas Islands*, by Emily C. Donaldson

Living with Oil and Coal: Resource Politics and Militarization in Northeast India, by Dolly Kikon

Caring for Glaciers: Land, Animals, and Humanity in the Himalayas, by Karine Gagné

Organic Sovereignties: Struggles over Farming in an Age of Free Trade, by Guntra A. Aistara

The Nature of Whiteness: Race, Animals, and Nation in Zimbabwe, by Yuka Suzuki

Forests Are Gold: Trees, People, and Environmental Rule in Vietnam, by Pamela D. McElwee

Conjuring Property: Speculation and Environmental Futures in the Brazilian Amazon, by Jeremy M. Campbell

Andean Waterways: Resource Politics in Highland Peru, by Mattias Borg Rasmussen

Puer Tea: Ancient Caravans and Urban Chic, by Jinghong Zhang

Enclosed: Conservation, Cattle, and Commerce among the Q'eqchi' Maya Lowlanders, by Liza Grandia

Forests of Identity: Society, Ethnicity, and Stereotypes in the Congo River Basin, by Stephanie Rupp

Tahiti Beyond the Postcard: Power, Place, and Everyday Life, by Miriam Kahn

Wild Sardinia: Indigeneity and the Global Dreamtimes of Environmentalism, by Tracey Heatherington

Nature Protests: The End of Ecology in Slovakia, by Edward Snajdr

Forest Guardians, Forest Destroyers: The Politics of Environmental Knowledge in Northern Thailand, by Tim Forsyth and Andrew Walker

Being and Place among the Tlingit, by Thomas F. Thornton

Tropics and the Traveling Gaze: India, Landscape, and Science, 1800–1856, by David Arnold

Ecological Nationalisms: Nature, Livelihood, and Identities in South Asia, edited by Gunnel Cederlöf and K. Sivaramakrishnan

From Enslavement to Environmentalism: Politics on a Southern African Frontier, by David McDermott Hughes

Border Landscapes: The Politics of Akha Land Use in China and Thailand, by Janet C. Sturgeon

Property and Politics in Sabah, Malaysia: Native Struggles over Land Rights, by Amity A. Doolittle

The Earth's Blanket: Traditional Teachings for Sustainable Living, by Nancy Turner

The Kuhls of Kangra: Community-Managed Irrigation in the Western Himalaya, by Mark Baker

CPSIA information can be obtained
at www.ICGtesting.com
Printed in the USA
BVHW031542210720
584166BV00003B/13

9 780295 747590